EXTREME SURVIVAL

ALSO BY DR KENNETH KAMLER

Doctor on Everest

eXTREME SURVIVAL

A Doctor Explores the limits of

Human Endurance

DR KENNETH KAMLER

ROBINSON
London

Constable & Robinson Ltd
3 The Lanchesters
162 Fulham Palace Road
London W6 9ER
www.constablerobinson.com

First published in the USA by St Martin's Press, 2004

First published in the UK by Robinson,
an imprint of Constable & Robinson Ltd 2004

A copy of the British Library Cataloguing in Publication data is available from the British Library

ISBN 1-84119-879-X
ISBN 978-1-84119-879-8

Printed and bound in the EU

3 5 7 9 10 8 6 4 2

For my parents, Ethel and Willie
('Burro and Potoff').
They never entered an extreme environment,
but by their examples
I learned what willpower
and positive thinking can achieve,
and what devotion and
courage are all about.

CONTENTS

PROLOGUE
In Extremis 1
JUNGLE
The Most Competitive Arena on Earth 19
HIGH SEAS
The Moving Wilderness 91
DESERT
The Marathon of the Sands 135
UNDERWATER
The Pull of the Deep 171
HIGH ALTITUDE
In the Kingdom of the Gods 201
OUTER SPACE
Going Over the Edge 259
CONCLUSION
The Will to Survive 301

Notes and Bibliography 321
Acknowledgements 329
Index 337

PROLOGUE

IN EXTREMIS

IF THE CHANTING STOPPED, my patient would die. I was sure of it, as sure as anyone can be in a freezing tent on the highest mountain in the world, breathing only half the air there is at sea level. Pasang should have died hours ago. He was lying in a sleeping bag on a makeshift stone table – a pile of rocks arranged as evenly as possible to form an elevated platform. The cracks between the rocks were mortared with frozen urine, Pasang having become incontinent before I could get a catheter into his bladder. The propane lantern illuminating the tent cast a sickly yellow glow across his bloated face, pocked with blue ecchymotic patches where blood had accumulated and then dried under his skin. His eyes were swollen shut. I had seen healthier-looking patients die faster in intensive care units.

Gathered around the rock platform were Sherpas, mountain men like Pasang, who had come to pray. Though they were facing toward him, eyes wide open, they didn't seem to see him. Their lips were moving in response to a monotonous chant arising from deep within them, as if conforming to the sound being produced rather than creating it. More chanting came from outside – a disembodied chorus from tents around the camp where other Sherpas were keeping vigil. In the stillness of the night the effect was powerful, primal and unnerving: a quadraphonic rumble emanating from within the mountain itself.

While crossing an aluminium ladder placed horizontally to

bridge a crevasse in the ice, Pasang had slipped and fallen 24 metres, wedging himself headfirst between the narrowing walls of ice at the bottom. Working quickly, a combined team of Spanish and Nepalese climbers managed to get a rope around his waist and hoist him back up to the surface. But by then he had been refrigerated upside down for nearly half an hour. The climbers radioed me at base camp to explain what had happened and ask for advice, but diagnosis and treatment over a two-way radio isn't easy. Preliminary reports were being transmitted to me in clipped phrases by excited climbers with strong accents and weak radios. I didn't have much to go on and wasn't too sure what to prepare for.

Radio in hand, I stepped outside my tent into the subzero temperature. Looking halfway up the massive 610 metre frozen cascade of ice and snow that marks the majestic and awesome entrance to Mount Everest, I saw a cluster of black dots moving against the frozen white surface. Despite the static, I could make out that Pasang was bleeding from his nose, but conscious and even able to walk. They were bringing him to me at base camp. The dots started to move, slowly conforming to the outline of each gigantic ice feature they crossed as they worked their way down the slope. Then they stopped. The radio crackled on again. Pasang had started to stumble badly. He collapsed, and was now unconscious. They were trying to secure him to a ladder, similar to the one from which he had fallen, to use as a litter. Now they would have to lower him slowly and laboriously, relying on a combination of clever rope work and brute strength. In the hours it would take for my patient to arrive, I had only a few facts with which to make a diagnosis, but plenty of time to think about his treatment.

A head-on collision of this magnitude, followed by loss of consciousness, would likely be fatal even if it occurred inside the kind of well-equipped emergency room in which I often worked in New York City. But I was at base camp on Mount Everest, 5,352 metres above sea level, separated from the nearest hospital by some of the most remote and rugged terrain to be found anywhere in the world. Evacuation by air would be entirely dependent on the weather (usually lousy), the condition of the helicopter (often grounded for lack of parts) and the availability of the pilot (frequently away). If Pasang made it down to base camp alive, I would do everything I could to keep him that way until we could get him out. I wasn't sure though, whether my efforts would make any difference at all. High-impact head injuries were the accidents I feared most; they were the ones for which I could do the least.

The brain is delicate and unforgiving. To protect this most vital organ, the body shields it inside a rigid case. The skull is very effective in protecting the brain against incidental trauma, and may even stand up to the swipe of a bear paw, but it is not strong enough to withstand a 24 metre headfirst dive into a crevasse. Pasang had most likely fractured his skull; the nosebleed was probably blood flowing downstream from the broken bone. In these cases, however, it is not the fractured skull itself that causes the most injury; as long as fragments aren't propelled inward, the breaking bone absorbs the shock and the brain survives. Many fatal head traumas don't even break the skull. Pasang might have a fracture, but he was showing signs of an injury even more insidious and deadly.

Skulls have disadvantages. While a blow to the head may not break the bone, it can tear the blood vessels that wrap around the brain. The blood, leaking out but encased in an unyielding skull, has no place to go but inward, collecting into what is called a haematoma. Held in by the tough lining of the skull – the dura – the blood compresses the much softer brain tissue, forming a subdural haematoma. The classic symptom is a temporary recovery right after the initial shock, as Pasang had shown when he had, at first, been able to walk on his own. Meanwhile, the bleeding continues to fill the confined space until the pressure causes a loss of coordination and unconsciousness, like Pasang had next experienced. Further brain compression, though often due to a relatively small amount of blood loss, eventually shuts down vital functions and the patient dies. Like Pasang?

Pasang did make it down alive, though just barely, and now I was watching him die. The relentless pressure was starting to affect the electrical circuits deep down, the ones that send the pacing signals for breathing and heartbeat. Respiration and pulse were slowing. Less oxygen was getting to his brain, causing the suffering tissues to swell and take up more space, further accelerating the pressure. Pasang's body was about to shut down, inexorably following the rules of physiology gone bad.

Pasang was slipping away but, strangely, I was calmed and pacified by the hypnotic chanting that was engulfing us both. I felt as a drowning man must feel when, at the very last, the cold water makes his body seem warm and numb. Mechanically, yet dreamily, I administered oxygen through a face mask and fluid through an intravenous line; I was even unaware that I had changed the IV bag until I saw that a fresh one was hanging there.

My efforts didn't seem to be having much effect. Holding his wrist with my fingertips, I felt a weak, thready pulse, out of sync with the chanting. I was sure the monotonous sound was reaching my patient's ears and wondered what effect it could be having on his brain. Might it be possible that over the centuries, through a combination of natural selection, experimentation and lucky coincidence, the Sherpas had developed a method for matching the pitch of their chanting with the natural frequency created by the vibration of their brain waves? Doing so would set up a harmonic, a sound whose preciseness of pitch causes an object to resonate, multiplying the energy input and creating an effect far more potent than the stimulus. Applied to the brain, this effect might even be powerful enough to reverse a shutdown.

I have learned not to dismiss this kind of possibility. No course in medical school taught me the proper mixture of oxygen, IV fluids and Tibetan chants to treat a subdural haematoma in below-zero temperatures on a 5 km-high glacier. Nor do I see the problem very often as a surgeon in New York, specializing in microsurgery of the hand. I learned it from my own experiences, and from those of others, in a vast array of circumstances – treating patients in the jungle, in the desert, on the ocean, under the sea, atop the mountains and into outer space – all the places where the human body confronts an extreme and unforgiving environment.

Millennia ago the earliest explorers, at sea in wooden boats or on foot in deserts and jungles, carried with them the same ancient mysterious device that scuba divers and high-altitude mountaineers use today. It was by far the most complicated yet most reliable piece of equipment aboard *Apollo 11* when it landed on the moon. In the entire universe, no system more complex has ever been discovered than the human body.

It comes in two standard models, which have spread out over the earth in billions of copies. Though some parts have been modified to work effectively in varying climates and terrain, the basic design has remained relatively unchanged for several aeons. Nevertheless, no one would claim to understand how it works, least of all doctors like me who have seen it function in harmony and in chaos.

Human beings are both tough and fragile. They have populated the earth by adapting to its various environments, but thrive only on the land, in temperatures between freezing and 38°C, and mostly at heights below 2 km or so. The body's own physiological constraints confine it to less than one-fifth of the earth's surface; beyond that, the environment is too extreme for an organism that needs food and water daily, oxygen by the minute and heat constantly. The few million humans who live in borderline environments don't thrive, but they do survive at the edge of their own physiology. Like Pasang, they live on the periphery of habitable land in regions such as the Himalayas, the Amazon, the Arctic or the Sahara – regions that can sustain life, though only barely. The borders of these realms are defined by connecting the points at which their inhabitants' body defences can no longer match the insults of the environment.

No animal in its right mind ever intentionally puts itself in danger by going somewhere it doesn't belong. Human beings, on the other hand, are controlled by brains whose emotional and spiritual imperatives can override the survival instinct. Humans have always had an insatiable drive to explore. Now, with the accumulated wisdom of countless generations, we have developed the technology to cross barriers that had previously contained us for hundreds of thousands of years. The combination of wanderlust and technology has given us the temerity to believe we can take on the most extreme environments on earth and not just survive but adapt to them. Human bodies, however, are far more fragile than we'd like to admit. If our protection breaks down, we die easily.

I am a doctor on many scientific expeditions to the most remote regions of the world. My patients are the people who live on the edge of survival and beyond. I practise medicine in environments and in situations incompatible with life, often treating conditions I have never seen before, and sometimes never even imagined. The operation of the human body is mysterious enough under normal conditions; when subjected to the full power of an inhospitable environment, it can become completely incomprehensible. It's not easy to fix something when you don't even understand how it works, yet on these expeditions it has been my responsibility to try to counter the effects of environmental insults on human physiology. When things go wrong in extreme locations – when a mountaineer develops pulmonary oedema, when a diver gets decompression sickness, when a body part is severed in a remote corner of the jungle and when a man tumbles headfirst into a crevasse – all eyes turn to me. There is no adequate repair manual to consult; my only tools are those I have brought with me. There is often no reasonably safe place to work, or even to think, and the problem has to be fixed immediately. So I have learned to make up some rules as I go along, often aware that I am not just waiting for help to arrive – help isn't coming. At least not from outside. Sometimes help does appear, arising from within – from a patient's deep-seated will to survive.

I practise medicine in places where I don't belong, often where no one belongs, because I never lost the childhood instinct to explore. When I was eight, I spotted a book in my house called *Annapurna*. An odd title, I thought. I couldn't imagine what it was about, so I climbed up my father's bookshelf and pulled it down. It turned out to be the classic tale of the ascent of what was, at that time, the highest mountain ever climbed. The book opened up to me a world I never knew existed, impossibly far away from my apartment house in the Bronx. The idea of exploring hidden worlds took hold of me and never let go.

Growing up in New York, it's not easy to climb mountains, but I discovered that with a microscope, I could explore vast mysterious worlds without ever leaving my room. That started me on the path to explore the ultimate unknown – the human body. I became a doctor, taking a residency in orthopaedic surgery and eventually a fellowship in microsurgery, moving along the prescribed route toward becoming a member of the medical establishment. But I always held on to the dreams of adventure that *Annapurna* had given me. I learned to hike, canoe, sail and scuba dive so I could reach exotic places on my own terms. Finally, one day I called a climbing school in New Hampshire and signed up for lessons. My instructor was an itinerant ex-Green Beret from North Dakota. I assumed he and I would have nothing in common. He turned out to be an intelligent, sensitive soul who shared my enthusiasm for adventure – though his path in life had given him far more of it than mine had. We hit it off immediately.

Six months later he asked me if I wanted to join his team to climb mountains in Peru. He enjoyed my company, he said, and liked the idea of having a doctor along. I was immediately tempted by this opportunity to combine my medical training with my passion for adventure. I was curious about the challenges that the human body in general – and my body in particular – could meet. For a single guy with no steady job, like my Green Beret friend, pulling up stakes and going to Peru for a month was easy. I, on the other hand, was a surgeon attached to a wife and a hospital. Knowing my lifelong desire to climb, my wife supported me enthusiastically while admitting she would have preferred it if I hadn't wanted to go. Taking a month off meant rearranging my schedule to ensure that there would be adequate coverage for my patients. As chief resident, I could make all the arrangements, though I didn't know how my colleagues would react to my disappearance. But I was determined to go climbing, even if it meant the risk of getting fired when I returned. Being a surgeon, I told myself, is what I do, not who I am. I didn't want to become a prisoner of my profession.

The bigger the mountain, the more planning required to climb it. The mountains our team would attempt in Peru were nearly 6,096 metres high. Preparations were divided among the team members according to their interests and skills; the responsibility for all the 'doctor stuff' fell to me. To the other team members, this obviously meant preparing a list of all the medical supplies we might need. Less obvious to them, but striking to me, it meant I'd have to know how to use these supplies, some for conditions I'd never treated or even seen. Motivated by the fear of being found wanting as a teammate lay dying, I was determined to start at the beginning and not miss a detail. I noted down every problem that might befall a high-altitude climber from head to toe – fractured skull to athlete's foot, and everything in between. I read about every condition and carefully listed each medicine or supply I would need to treat it. If I didn't bring it myself, I wouldn't have it and there would be no way of obtaining it. I even wrote out treatments on small squares of paper that I would keep in my pocket in case my mind went blank in an emergency. In short, I went to Peru thoroughly prepared and somewhat confident but nevertheless hoping for a trauma-free adventure. I didn't want to be put to the test.

With that thought in the back of my mind, I sat on climbing gear in the back of an open truck and watched the Peruvian countryside roll by. We were riding up a high mountain pass on our way to Taqurahu, the 5,791 metre peak that we intended to climb. Another open truck, even more rickety than ours, was coming from the other direction filled with Indian villagers on their way to market. As we watched in disbelief, the truck teetered on the edge of the road and then toppled over, tossing out people and animals as it tumbled down the slope.

My first test was going to be at a full-blown disaster. It was like a bad dream. I forced myself to keep cool on the outside, because if the doctor looked nervous, everyone would get nervous. The two minutes it took for our truck to reach the accident scene was enough time for me to calm down on the inside by focusing on the likely injuries. I ran through the treatment steps in my head. I remembered the notes in my front pocket.

People, livestock and baggage were strewn all over the hillside, which sloped down to a ravine where the truck was lying on its side. I paused at the top to let the first frightening impression pass by, then took an analytical look, noting who was groaning, who was only moaning, and who was bleeding. No one seemed to be dying. I reduced the chaotic scene to a series of problems I would handle one by one. Gradually it became clear that although I was high up in the Andes, I was facing injuries not unlike those I would find in any hospital's emergency room. I injected a little girl with an anaesthetic, then set her fractured forearm. My Green Beret climbing teacher was a resourceful assistant; he splinted the arm with a wooden slat he broke off from the overturned truck. I started an intravenous line on a farmer who looked ready to faint. There was one concussion, one blunt abdominal trauma and several other more minor injuries. Once I realized there was no patient I would not be able to treat, my confidence, which at first had required some effort to project, started flowing naturally. Though they spoke only Quechua, my patients and I communicated easily in the universal language of patient and doctor. I didn't once think about the little treatment notes in my pocket, though having them there may have helped me subconsciously. As I finished cleaning and sewing the last laceration, one of my patients came back with a goat, still stunned and bleeding from the neck. Buoyed up by how well things had gone, I sewed up the goat.

According to our map, a *clínica* was a day's ride away, so we unloaded our truck and laid in all the villagers for the long trip. One of my teammates came with me; the others just set up camp where we were and waited for us to return. We fully expected to drop off our patients and make a quick U-turn to the mountain. The *clínica*, however, turned out to be a cinder-block room with no medical supplies, run by a doctor who seemed capable only of handing out birth-control pills. Although the patients were stabilized, he begged me not to leave until medical transportation arrived from Lima. With the still vivid memory of the fear that can be provoked in a doctor faced with a medical challenge, I had empathy for my colleague and stayed the night.

The rescue was publicized in Peruvian newspapers and on the radio, making me something of a local hero. We eventually did get back to climb Taqurahu, but reaching the top wasn't nearly as exhilarating to me as treating the accident victims had been. I had risen to both challenges, and yet, in my heart, taking care of the villagers had pleased me more. I was interested in the earth's extremes, but succeeding at extreme medicine had been the higher summit.

When I returned to New York, I was relieved to find that my hospital supervisors, though they couldn't admit it officially, seemed to admire what I had done. Only a token punishment was imposed for my absence. Even had the consequences been more severe, however, I knew beyond question that I had made the right decision. I was determined to open even wider the door to that other world.

My exploits in Peru came to the attention of The Explorers

Club in New York, and not long after my return I was invited to join that venerable group of seasoned world explorers. For my first meeting I was asked to come with one good idea. Knowing that its membership must possess a unique and wide-ranging collection of medical experiences, I suggested that someone first collect and then synthesize the information, to create a fund of knowledge on medicine in extreme environments. The idea met with approval and, as often happens to someone with an idea at a meeting, I was unanimously assigned the task of bringing it about.

At first it seemed presumptuous that a specialist in microsurgery such as myself should aspire to become an expert on extreme medicine. I felt like an outsider looking in. As I delved into medical journals, however, I found that there were few articles to read; fewer still were worth clipping and underlining. What little information existed was often vague, impractical or contradictory. No doctor, I realized, could truly master such a disparate and random collection of far-flung maladies, but, within the Club, I was very quickly perceived as the repository and the source for information.

My Explorers Club comrades were eager to share their experiences with me, and I soon had a collection of practical advice on treating medical problems in places and settings I never could have imagined. Perhaps I really did have a unique position from which to practise extreme medicine. Explorers heading for every part of the globe routinely began coming to me for advice, often coupled with invitations to join their expeditions. For someone who a year earlier didn't even personally know a 'real' explorer, every offer seemed too good to refuse but by this time I had left the hospital and was in solo private practise. No doctor I knew of crossed routinely between the worlds of exploration and medicine, and there are good reasons why. Setting up a practise involves risk and investment. I was proposing to be away for long periods of time. That would mean a loss of continuity for my referring doctors, not to mention a loss of income. The effect on my practise would be unpredictable, but taking risks and facing the unknown are what explorers are supposed to do. The experiences would be worth far more than any acquisitions.

So I became a medical explorer, stepping out into the most extreme environments in the world eager to confront unexpected challenges. I paddled through the Amazon in a dugout canoe. Crossing the Arctic tundra, I tried to remember exactly how I was to record the migration route of a polar bear that was just then banging its lethal paw against the steel-reinforced window of my buggy. I had to stop taking notes for a fish survey while scuba diving in the Galapagos Islands when a shoal of hammerhead sharks passing above me obscured my light. On the Antarctic plateau, in a whiteout so severe I couldn't see my feet, I made my way back to my snowmobile only by managing to follow voice cues from the driver.

Wherever I was and whatever the circumstances, I was always the doctor, expected to treat whatever injury, or insult, a hostile environment might inflict on a fellow traveller – from frostbite to snake bite. If I didn't know what to do, I would rely on local lore or improvise. I took my expedition work very seriously and after a few years, wasn't quite sure any more which of my worlds was the 'other' one. Dr Kamler's adventure stories always circulated quickly around the medical community, often embroidered with more detail and infused with more drama than my original accounts. Increasingly I felt like an outsider in the traditional, medical world from which I had come. Sometimes my New York practise seemed the more alien environment. However, something very interesting happened. Doctors pinned down by heavy mortgage payments and high overheads sought me out for vicarious relief. They understood why I did what I did. They shared my curiosity about what happens to the human body at the limits of medicine and felt the same longing for adventure in places far from their waiting rooms.

The earth's harshest and least explored environments – above sea level, at any rate – tend to contain mountains. Many of the expeditions I have found most intriguing involved rugged mountaineering, making climbing proficiency a required skill for a doctor practising extreme medicine. I climbed in the Alps, the Andes and Antarctica, joining the tight community of high-altitude climbers. No matter what hemisphere I was in, I would run into the same people.

Several years ago, by word of mouth, I heard of an expedition being put together, sponsored in part by *National Geographic*, to study the tectonics of Mount Everest and measure its exact height using a laser telescope. The climbing would be difficult, but the research would be valuable, and the challenge of providing medical care would be as enormous as the mountain itself. My phone call to the expedition leader struck him, he later told me, as divine intervention. Though he had a clear scientific objective, adequate funding and supremely qualified climbers, he had not yet solved the problem of medical supervision. I would be the only team member he hadn't already climbed with and my climbing resumé was a little thin by his standards, but he was still eager to have me. He had been to Everest before and told me how quickly bad things can happen on big mountains, even to experienced people. He was right. Only one day after we arrived in base camp, with most of my supplies still in boxes, Pasang fell into a crevasse. I was yet again treating someone in a hostile, unforgiving environment and discovering – yet again – the body's enormous capacity for survival.

That discovery forms the core of this book. We are going on a journey into the most remote and dangerous regions of the world, and then continuing on into the bodies and minds of the people who are there, people for whom that environment is very real and very life-threatening. Some of the stories are journeys I have made myself, some are based upon the experiences of others: climbers, divers, sailors, explorers, astronauts, as well as ordinary people who found themselves in extraordinary circumstances. Woven throughout are observations and reflections on the evolutionary biology, physiology and psychology that combine to give humans the means to prevail, whether it is in an acute response to an attack from a force of nature, or a long-term adaptation to a chronic stress in the environment. We will cross the human threshold to see how the body works under normal load conditions, how it moves into overdrive when subjected to environmental insults and, finally, what happens when the body breaks down – overwhelmed by extreme environmental forces it was never designed to withstand.

The tried-and-true method for dealing with extreme environments is to avoid them. Apart from the occasional hapless wanderer, this approach has worked well for the human species over a few hundred thousand years. People who live on the edges of these no-man's-lands have adapted to their particular environmental stresses through natural selection, but it made no sense to venture any farther inside to a place offering no food or shelter. And there was no good way to get there, in any case. Ocean depths, remote deserts, high seas, dense jungles, tall mountains and outer space were all safely inaccessible, so the challenge to adapt to more extreme conditions simply did not exist. With gradual exposure over enough generations, humans demonstrate enormous adaptability. Without that exposure, such remote regions are deadly.

With no one willing or able to go there, forbidding environments long remained empty spots on the map. Advancing technology, however, has suddenly made these places accessible. Moreover, the peculiarities of modern civilization have made them alluring to explorers, scientists and adventurers. In an evolutionary instant the rules have changed, though the game – survival – remains the same. For example, the body 'understands' tiger bites; it doesn't 'understand' nitrogen bubbles. It knows the rules for tigers because it has been dealing with them for thousands of years. Through trial and error, natural selection has equipped humans with a complex and precise sequence of physiological responses, refined over countless generations, whose aims are to counteract a tiger's predation. First, the mind recognizes that the fast-approaching creature is dangerous; then it sends a signal to the body, instructing it to flee. Should the escape mechanism not work, the body prepares to defend itself. And if it gets injured, there is a preset sequence of healing cells and chemicals that can assess the damage and initiate repairs. The effort may prove unsuccessful, but the body understands the threat and has developed the tools that enable it to fight.

What happens to the same meticulously developed defences when nitrogen bubbles attack the body? Deep-sea diving has only been around for a few generations, and exposure, when it occurs, is anything but gradual. In the body of a diver, the bends, the onslaught of nitrogen bubbles in the bloodstream, is like an alien invader. Having never seen it before, and with no time to adapt, the body's response will be as chaotic and misdirected as planet Earth's reaction would be to an attack from Mars. The body undertakes equally chaotic responses to other unfamiliar enemies: the low air pressure on a Himalayan mountain or the high water pressure under the ocean; the constant daylight of a polar summer or the constant darkness and extreme cold of a polar winter; the relentless heat in an African desert; the weightlessness of outer space. Yet, as we will see, people who live on or near extreme environments – from jungle Indians to Sherpas, from Eskimos to Bedouins to South Sea pearl divers – have each developed specific adaptations to the environmental insults they confront every day. Evolution has moulded humans to fit along the various frontiers of survival, but even their special protection breaks down quickly if they push themselves beyond their borders.

Pasang should not have survived, but he did. Over the course of the night, his thready pulse strengthened, the swelling in his face receded and he opened his eyes. With the morning light, the chanting stopped and the spell was broken. Though I felt I had been watching the scene from afar, I was certain I had witnessed a healing force that was beyond medicine. The chanting had released an energy within Pasang, a will to live, and this had reversed his decline. A spiritual force had created a tangible effect, what a religious person would call a miracle. My medical training should have led me to explain his recovery in terms of nerve impulses and chemical reactions, but confronted with such incontrovertible testimony high on a Himalayan mountain, even a faithless man believes. There was no medical reason for Pasang to be alive. I realized then, that practising extreme medicine would sometimes mean witnessing, and working alongside, phenomena I might never understand.

A rescue helicopter was on its way and we had to get the patient ready to go. I busied myself with the details of an evacuation that, last night, I was sure wouldn't be necessary. Conscious now, though still only dimly aware of his surroundings, Pasang began protesting when his arms were tied to the stretcher. It was a good sign – although the words, I was told, were a Tibetan curse.

Every time I returned to Everest, I recalled the power of that event. The research with which I was involved expanded to include global positioning satellite beacons as well as lasers, and took five seasons to complete. With every new expedition, I became more familiar with Everest but never lost my respect for it. Though scientific instruments were placed on the summit, none of us ever felt as if the mountain had been conquered, or that it could be conquered. There were too many reminders of how dangerous a place it is, and how frail in comparison are those who dare to climb it.

In 1995, as I was ascending a steep, overhanging pitch, a Sherpa team member slipped on the slick ice directly above. I watched in horror as he fell past me, plummeting 914 metres – almost a kilometre – to his death. The following year I was again trying for the summit when a vicious two-day storm took the lives of eight of my climbing friends. I was the only doctor on the mountain and did what I could to help the survivors, but my team was powerless to save those who were lost and freezing in the snow. They were as much beyond help as if they had been lost in space – an analogy that stimulated NASA to try to apply some of their space-age technology to the problem of saving people lost in the wilderness.

In the year following that storm, I got a call from a NASA Commercial Space Centre that was funding a programme to field-test medical monitoring equipment under the most extreme conditions possible. If their equipment could be made to work on Everest, it would work anywhere on earth – or beyond. They had been looking around for an experienced doctor to serve as their 'chief high-altitude physician,' and my name kept coming up. Here again was an offer I could not refuse. Climbing Mount Everest is the ultimate test for a mountain climber, and bringing the world's most sophisticated medical care to the world's most remote environment would be the ultimate test for an extreme medicine doctor. I accepted the challenge for myself and for the friends I've lost in the mountains.

I returned to Everest, not despite the tragedy of 1996 and the others I've experienced, but because of them. I worked with scientists and engineers from the Massachusetts Institute of Technology, Yale University and the Defence Department to develop computer models and treatment protocols that we tested on Mount Everest. The same wireless body sensors NASA is developing for astronauts on a space station or on Mars were adapted to be worn by mountain climbers, or any other wilderness travellers. Sensors continuously transmitted heart rate, respiration, body temperature and other vital signs, as well as exact location, to us at base camp. We knew at all times whether and where someone needed rescue.

Once a sick or injured climber was brought to our medical tent, we supplemented the data with heart sounds, breath sounds, EKGs, sonograms, microscope slides and video images of the patient. The digitized information was sent via satellite through a live-TV hookup to Yale and Walter Reed Hospitals so medical experts there could radio back real-time treatment advice.

We had foreign climbers come to us at base camp asking to be treated for chronic conditions; the level of care we provided on the mountain was so much higher than anything they could receive at home. Had such a telemedicine system been in place during that fateful storm in 1996, we might have been able to save some of the people high up on the mountain who are still lying there.

I'm mindful of how far I've come since the days when I kept those little treatment papers in my front pocket. I've been to some of the most remote regions on earth, and I've had the rare privilege to practise the only form of medicine that mixes modern drugs with herbal cures, satellite signals with ancient chants and science with spirituality.

JUNGLE

THE MOST COMPETITIVE ARENA ON EARTH

THE BLACK, STAR-FILLED LAKE was such a perfect reflection of the night sky that my paddle made ripples through each constellation. With stars both above me and below, it would have been easy to imagine I was canoeing through space – had it not been for the glassy eyes that broke the water like closely set periscopes glowing jewel-red in the reflection of my flashlight. The eyes remained unblinking, motionless, but my strange light stirred up a hollow grunting sound from just below the surface. Many other pairs of rubies were scattered among the stars in the water. As we passed each of them, the same warning noise arose from below. In the humid silence, the grunts resonated over the lake, intensified by the darkness. I could barely see my companions in the front of the canoe, and no one spoke. Each of us was alone. Hours passed that way, or maybe it was only minutes.

Gradually I grew aware of light coming from above and behind me. Over the jagged horizon appeared an orange disc. I knew it was the moon, but I felt as if I were witnessing sunrise on some other planet. The stars dimmed; the sky and the water turned silver. An irregular black seam appeared around the edge of the lake, its top and bottom edges perfectly symmetrical: the silhouette of the surrounding treetops mirrored in the still water. In this bizarrely lit world filled with unfamiliar noises and inhabited by strange creatures, I was an alien.

I was invading the Amazon, a place crowded with more plants and animals than anywhere else in the world, and therefore the arena for the most unrelenting and fearsome battles for survival. I had come here to study one of the winners of that competition, a prehistoric relic so perfectly adapted to this environment that it has lived here, essentially unchanged, for half a million years. By contrast, I had been here a week, having arrived by plane and canoe from my home in New York City, an environment far more unnatural than the jungle, but my native habitat nonetheless. My lifetime of adaptation to many of its most prominent features, such as streets, electricity, advertising and air pollution, was irrelevant here. However, some of my acquired skills, particularly those in medicine and surgery, retained much of their value anywhere, even and especially here – in the Upper Amazon Basin of Ecuador, on a lake so remote it's a three-day canoe trip from the nearest town and reachable only during the rainy season, when water levels are high enough to make its shallow entrance channel navigable.

The pristine lake provides an ideal study area for a team of field biologists intent upon unravelling the secrets and survival strategies of plants and animals that can't live anywhere else on earth. It is also an ideal setup for those very plants and animals to prey upon the biologists who have come to study them. There's no mercy here for anyone whose bodies and skills have not been honed by natural selection and toughened by a lifetime in the jungle. Distracted by their work, the scientists knew they would be easy targets. I got a ticket on to this expedition because they needed a doctor who could at least try to repair the damage – anything from an itch to a cardiac arrest – that might be inflicted by the toxins, spines, stingers, claws and teeth that surrounded them. I had been eager to join the team; I wanted the challenge of treating exotic diseases in an exotic setting. Even more, I wanted the chance to become enmeshed in a tangle of life so densely interwoven that in the month we would be there I would only barely begin to glimpse its complexity.

Fully intertwined in this jungle is another well-adapted species, two examples of which were with me in the canoe on that starry night, visible now as silhouettes against the backdrop of moonlit water. Crouching low at the very front was Berullio, the builder and proud owner of our dugout canoe, the largest in the area. Behind him, sitting in the bow, his father-in-law, Antonio, was steering. Two non-native specimens were paddling in the stern: the expedition leader, a college professor from Connecticut, and myself, the expedition doctor. Antonio and Berullio are Cofan Indians, descendants of one of many small groups of humans who have adapted to the Amazon after thousands of years of harsh natural selection, tempered by countless generations of accumulated wisdom and practical experience.

Again the silence on the lake was broken by a grunting noise. This time, however, Antonio had produced it. He was immediately answered by a deeper, louder grunt coming from near the shore. Berullio signalled us to paddle in that direction. Antonio and our target continued their conversation while we made our stealthy approach across the dark lake. We swung into a cove. Suddenly we saw pairs of red eyes everywhere: some glowed at us from among the submerged trees and tangled grasses that ringed the shore; others were moving along atop black shadows that skimmed the silver water surrounding us. We heard many more answering calls now. Berullio and Antonio remained focused on one pair of eyes, wider apart than any of the others we had seen. They were like red lights mounted on the end of a thick black log. The distance between the eyes is proportional to the size of the animal, so even I realized this creature had to be massive.

The boat glided straight on toward the shore, the college professor and I in the stern, paddling gently, and Antonio making fine corrections with his paddle in the bow. He put his light out; we did the same. Berullio's light, fastened to a band around his head, remained on; any other light would be distracting for what he was about to do.

The shoreline trees loomed larger and lower until they were

hanging over us, their twisted branches and dangling vines lit up by Berullio's headlight. The beam was bouncing off the water in a moving cone of light that illuminated the shallows of water hyacinths and elephant grass as well as the interlocking branches of the canopy overhead. Berullio tightened his crouch, to wedge himself a bit more inside the tip of the bow, then carefully leaned over the left side. In his left hand he held a lasso that was fed through a hollow bamboo tube that he held in his right hand. We were moving the canoe very slowly now – hardly at all. The eyes lay dead ahead. Antonio tapped the water with his paddle. The canoe turned and the eyes swung left, now almost even with the bow. I could make out a black head, like an island surrounded by the silver water. Berullio lowered the bottom of the lasso into the water with a slight motion of his hand, making no sound at all. He placed it just in front of something else protruding from the water: two large holes that I suddenly realized were nostrils, connected underwater to the rest of the head that was still about a metre farther back.

Berullio held the rope steady, letting the momentum of the boat slowly carry the lasso under the mouth and jaws. Then, in one quick move, he lifted the lasso over the creature's head, gripped the bamboo pole and jerked the loose ends of the rope with his right hand to tighten the noose. The rope caught on a water hyacinth and slapped against the head instead of slipping over it. The animal, until then entirely motionless and seemingly oblivious, suddenly reared its head back, lifted its upper body out of the water and gave the canoe a ferocious kick, lurching it sideways. I gripped the sides of the rocking boat and felt lake water spilling over my knuckles. Berullio's hand got caught in the rope. The beast thrashed its jaws as both it and Berullio tried to disentangle themselves. Berullio's headlight ran helter-skelter over the animal, giving us quick flashes of rows of sharp white teeth lining a long pink throat. My heart was pounding. I didn't know what to do and there was no time to react anyway. In another second came a loud splash. The animal disappeared under the water, tugging Berullio's arm down with it. Some bubbles rose to the surface, and then Berullio calmly withdrew his arm from the water, holding an empty lasso. I felt immeasurable relief: no mauled arm, not even a cut.

The crocodile had escaped. Antonio said that judging by the space between its eyes (about 10 cm) the animal was approximately 5 metres long. I could easily believe that. I had seen for myself that the distance just from the nose to the eyes was about a metre. Different species of crocodile have different eye colour; on our three-day approach along the river, I had seen only green and orange eyes. This lake was filled with red eyes – the colour of the black caiman. Its eyes are actually colourless. The red is produced by our lights reflecting off the blood vessels inside. The black caiman is a fearsome but strikingly beautiful animal with a soft leather skin, black with yellow-white stripes.

'I should have shot it,' Antonio said ruefully. 'A good hide and a lot of meat.'

He couldn't quite understand why a group of gringos was asking him to catch these animals and bring them to camp, where we would perform the frivolous rituals of weighing and measuring them, counting their teeth and painting numbers on their heads with nail polish. Worst of all, when we finished we would ask Antonio to drop them back in the water where he found them – what a waste. But the pay was good, and cash was hard to come by in the jungle, especially now that poaching was illegal.

'Black caiman skins used to bring good money,' Antonio told us. 'Not the other crocodiles – they have little bones in their skin. At the market no one wanted them. But the black ones make beautiful leather.'

I suddenly realized why, on our three-day river approach, we had seen only the orange and green eyes of the other crocodile species and came across red eyes only after we had entered the nearly inaccessible lake.

'Not so many black caimans any more,' I said to Antonio.

'No, hard to find now,' he agreed.

I wondered whether Antonio's poaching days really were

over or whether he'd be back here as soon as our expedition left. I also wondered about the fate of the other crocodiles. A new tanning process had been developed that could dissolve small bones embedded in hides. The local Indians would find out about it soon enough.

We continued our hunt around the lake, targeting a second pair of eyes (less far apart than the first pair) and edging silently toward it. This time Berullio got the noose tightly around the creature's head. Antonio jumped up and coiled a rope around the snout. Then they put their forearms in the water under the animal and forklifted it into the canoe. I felt its power as it came on board. Though I was seated safely at the back of the boat, a force of nature had entered my personal space.

The crocodile wasn't too cooperative, so it was roped further. The forelegs were tied to the hind legs and then to the tail, to stop it from flailing. An Amazon rodeo. Once the animal was subdued, it was dropped in the middle of the boat, behind them and in front of us. The terrifying and majestic ruler of the lake had been transformed into a pitiable spectacle in the bottom of a canoe. I stared at it. Antonio made a sharp whispered noise to catch my attention and then swung his head to one side, indicating the direction in which he wanted me to paddle.

Of the dozen pairs of eyes we targeted that night, Antonio and Berullio lassoed six. None was longer than 2 metres, but as the night wore on we accumulated quite a pile. Our boat was very big – about 8 metres long and 1 metre wide at the beam – but it was still a dugout canoe, hollowed out from a single tree. It had no keel and was very tippy. There were numerous cracks in the side walls. I had been told that the boat was five years old.

'How long do these canoes last, Berullio?'

'About five years.'

Each time we took on another unwilling passenger, the professor and I grabbed the sides of the canoe to steady it as it rolled from side to side and sank lower in the lake. With the water level only a few millimetres below the edge, the cracks became leaks. I was able to staunch one large one by wrapping a plastic bag over a pencil and jamming it in as a plug. But water continued to drip, supplemented dramatically each time a vigorously protesting caiman brought some of the lake in with it. The puddle on the bottom was good for our captives but bad for my feet; my waterlogged boots had remained immersed for hours. Finally I had enough. Far more insouciantly than I would have thought possible one week earlier, I took off my boots and rested my bare feet on one of the crocodiles.

Just before dawn, the temperature on the lake dropped. I had been cold and wet even before it started to drizzle, so despite our ponchos the professor and I were pretty uncomfortable. The Indians in front had no rain gear, but their enthusiasm seemed undampened. They kept sweeping their lights across the water. The caimans, however, were getting harder and harder to spot. Antonio explained that raindrops annoy the creatures, so they were closing their eyes. As the rain got heavier, the red lights all went out. We turned our canoe back toward camp.

Sunrise comes suddenly at the equator. As we paddled back, the rain stopped and the sky evolved in rapid sequence from silver to puce, mauve, pink, orange and then blue. A huge ball of fire made its dramatic entrance over the trees. I was starting to feel drier already, but was still looking forward to a change of clothes, breakfast and a good morning's sleep. Tonight our team had come close to experiencing its first crocodile bite, but having seen Berullio deftly extricate his arm, I felt a little more relaxed. I was beginning to think I might get through this expedition without treating any major lacerations.

With the dawn and the proximity to shore, the jagged silhouette that had rimmed the lake all night revealed itself to be a tangled wreath – a 46-metre interlocking wall in every imaginable shade of green. From a distance, individual trees were indistinguishable except for the kapoks that punched through the top of the canopy and rose an additional 30 metres before spreading out into huge crowns. They looked like gigantic heads of broccoli – dark green against the bright blue sky.

As we got closer in I could see the detail of the shore, but nowhere could I see dry land. The lake ended in a succession of plants, their stalks protruding from the shallow waterlike weeds with floppy leaves on top, called elephant ears. Mixed among these were delicate-looking purple flowers growing on flat, floating leaves – the water hyacinths that nearly cost Berullio his arm. Looming behind them were the first rows of trees, actually growing out of the water, and above them a slope of muck that led to higher ground. The growth there was too thick and too dark to see anything within.

Improbably, a little boy popped out from the maze of foliage. He was waving his right arm and holding up his left hand, which was dangling forward and to the side, a position we surgeons call the 'wounded paw'. Hands can't fall into that odd angle unless they've sustained major damage. The boy needed help. It was as if my beeper had gone off, transforming me from dreamy jungle gazer to doctor on alert.

The boy came to the top of the mud slope and shouted something to Berullio. He couldn't easily get down through the ooze, and our canoe couldn't penetrate any closer to shore. There was a short, excited conversation I couldn't make out, and then the boy disappeared back into the vegetation. The only word I had caught was 'doctoro'.

Antonio made a chopping motion with his right hand against his left wrist. 'Machete,' he said.

'When?' I asked.

'This morning,' he replied. 'He'll meet us at the village.'

Being in charge of our logistics, Antonio knew that my medical box – as well as most of our other equipment – had not yet been transported to the lake but was still at the village. The rest of the team would be there this morning, loading supplies. It would be the best place to treat the injured boy.

Antonio had decided this without even consulting the *doctoro*. I could have been insulted, but that would have been silly. This was the jungle, and here Antonio was in charge.

Before the boy appeared, we had been on our way to our

lake encampment. The village was located on the river. We changed direction, heading toward the mouth of the lake. Antonio and Berullio conversed about something, told little jokes and in general showed no signs of urgency. This was one more trial of the jungle. The tough would survive, the weak would not. We paddled on at a pace no faster than before.

Exiting the lake, our canoe entered a current of muddy water. We paddled upriver around submerged logs and twisted branches to a wide mud beach. Above the beach, a bank of mud sloped upward to a clearing in which I could just make out the tops of two huts. Other canoes from our expedition had already arrived and were taking on some supplies, getting ready to set off for our research camp on the lake. A dozen kids ran up and down the beach, enjoying all the excitement.

I stepped out of the canoe into soft mud and sank to my ankle. I made a second step and went in up to my boot top. The thought of the third step was too much even for the Indians watching from the shore. They signalled me to wait, then hauled a canoe across the beach and pushed one end toward me. I stepped into the canoe, walked the length of it and then stepped out on to firmer grey mud. The *doctoro* had made his entrance.

My patient was already waiting. He came directly up to me on the beach and silently held his left arm up for me to examine. He looked to be about nine, meaning he was probably seven. Kids mature fast in the Amazon. I didn't know any words in Cofan, the local language, and he didn't speak anything else, but an easy smile and a confident manner don't require translation. Still, I couldn't tell if we were communicating. His eyes were open wide yet too dark to read, and his face remained expressionless. I took his wounded hand in both of mine.

The wrist was caked in mud and dried blood, the hand drooping forward. I wanted to get a look at the cut, but I knew that once I removed the mud it would start bleeding again. As we stood there, surrounded by a crowd of adults and children of all ages, I spotted what looked like a clean rag in the sand and walked over to pick it up. All eyes followed my every step.

The audience was far more intrigued by the Western doctor than by the injury he was treating. Cuts, they've seen before. I started wiping the wound. The boy watched carefully, still without expression, even when the plug of mud fell away in one lump and a pulse of blood spurted out. By the second pulse I could see that the cut was very deep, running straight into the wrist from the thumb side. I used the rag for compression, placing it where the mud had been. From the liveliness of the bleeding and the dropped position of the wrist and thumb, I knew an artery and some tendons were cut. It was likely that the nerve running between them had been cut as well.

'How did he do this?' I asked Sebastián, our guide, who spoke Cofan, English and Spanish and often needed all three to keep the group together.

'He was in his father's canoe this morning, clearing a channel. The water level was low and the canoe was caught on some submerged grasses. He reached underneath, grabbed the stalks with his left hand and chopped at them with the machete in his right hand. He missed.'

Though this wasn't the kind of patient history I get too often in New York, the type of injury was certainly familiar. I had treated one just like it only a few months before. A teenage boy cut into his right wrist in what I initially assumed was a suicide attempt. His psychiatrist pointed out that the boy was right-handed, however, and that had it been a suicide attempt, he would have used his right hand to slash his left wrist. In fact, the boy had been trying to cut off his right hand as punishment for masturbating. I spent hours and hours in an operating room, using a microscope to repair his hand so that he could masturbate again. That was the kind of incident that could only happen in a modern, highly developed country, someplace you couldn't get to from here. Perhaps more than anything else up to now, the thought made me realize how far I had travelled.

'He is going to need a full-blown operation,' I told Sebastián, 'but I think we can do it here. If not, he is going to permanently lose a lot of the use of his hand.'

'Of course,' Sebastián replied. 'What do you need?'

We were still standing on the grey mud beach, and the morning sun was getting intense. I suggested we move up to the small clearing, where the footing was better and where we would have some shade. Sebastián spoke to Berullio, who found an empty oil drum, placed it near another one already there, then returned with a wooden plank, which he laid across them to give me a narrow table. The drums were stuck deep into the mud and the plank was stable, but it was only about a metre off the ground.

The boy sat straddling the plank, leaned forward, and extended his arm on to it, but it was too low for me to work on. Bill Jahoda, the college professor and our expedition leader, saw the problem right away. He volunteered to sit on the other side of the plank to support the boy's arm on his legs. That raised the wrist another 30 cm off the ground. We were at the edge of the clearing where the slope was still steep enough so that if I sat on my knees in the mud in front of the table I was at about the right height. It wasn't my three-way tilt chair with padded armrests that I was used to at my hospital, but the arrangement suited me perfectly.

By now, a crowd of about thirty people (more than the entire population of the village) had collected around us. Sebastián attempted to disperse them by announcing that our crew was down on the beach giving out oranges. The adults left, but the little kids all remained to watch. They didn't care what Sebastián said.

I kept my surgical instruments and most of my supplies in a large fishing tackle box wrapped in a waterproof plastic bag in one of the canoes. A crewman went down to get it while another took two pails to fill with rainwater from an open drum behind one of the huts. They spread out the plastic bag on the mud and placed the box and the two pails on it as if they were setting up for a picnic. I opened up my medical kit and laid out my instruments and supplies in the order I expected to use them – gauze pads, retractors, haemostat clamps, cautery, scalpel, forceps, scissors, needle holder. This is a job usually done by my scrub nurse, so I paused to visualize every contingency and make sure I had everything I needed. My immediate environment, at least, was becoming familiar. I was in the OR now. Up to my knees in mud.

I put on sterile gloves and used a Betadine scrub to paint the arm and hand, as well as part of Bill's hand and quite a bit of his pants, as he was unflinchingly holding the arm across his legs. Then I laid a sterile drape under the boy's hand and across Bill's lap, tucking it into his pants so it wouldn't slide off.

Sensation to the hand and forearm is supplied by three major nerves, two of which ran into the territory of the injury. To provide anaesthesia, I would have to block both, using lidocaine (the same medicine a dentist uses to numb teeth). The injections were exactly what I would be giving in a regular operating room, but here they almost seemed unnecessary. An injury of this magnitude should cause intense pain, yet the boy made no complaint. In fact, I hadn't heard him utter a word since I met him. He was calmer than I was.

All doctors have seen examples of patients with major injuries who feel no pain. Pain is an alarm that alerts the conscious mind to bodily injury, motivating it to do something to correct the condition. When the damage is massive the problem is obvious; no alarm is necessary. A flood of pain impulses proportionate to the injury would serve no purpose, and in fact would interfere with the signals the brain is trying to generate to coordinate an effective response. The more desperate the situation, the more critical it is that the pain signals be squelched. It is a universal survival mechanism, one activated more frequently and more forcefully as environments become more extreme.

As I drew up the lidocaine into a syringe, I told Sebastián to tell the boy that this would hurt but that he would feel no pain afterward. The words seemed to me a rote formality. This boy was the most stoic patient I had ever seen. Berullio squatted behind the boy and held his shoulder. When the needle went in, the boy screamed and started to cry. I was startled. The machete wound had engendered only carefully calculated, purposeful responses, yet a simple needle stick nearly brought on hysteria. I shouldn't have been surprised. The reaction made sense from a survival viewpoint. The situation was under control. The boy could now afford the luxury of feeling pain again, and this was just the sort of injury for which an alarm is needed. The sensation is equivalent to being stuck by a thorn, a spine or a splinter – the kind of trauma that if ignored might lead to sustained bleeding, poisoning or overwhelming infection. Crying would not have been useful while he was in survival mode, but now the boy was in a protective environment. He was surrounded by people who could offer him aid and solace if he made it known that he needed it. I was reminded of my daughter. One day while playing at home, she fell over a chair. I was home, but my wife was not. My daughter got up without a peep and continued her game. A moment later my wife came home, and my daughter ran to her, screaming in pain. Subconsciously she had judged that reacting for me wouldn't be worth much but that reacting for mommy would bring a much greater reward.

It was taking a few minutes for the nerve block to work. I had given the injection in the forearm, 'upstream' of the injury, and the lidocaine was slowly flowing down the nerves into the hand. While we waited, Berullio comforted the boy, wiping his tears and helping him blow his nose. With a start I realized that this was Berullio's son, a fact that had been common knowledge to everyone but me. His name was Hermanigildo. Antonio was his grandfather. I had been so tightly focused that I had been oblivious to any social nuances around me. The father–son interaction that I was seeing now was what you would expect in any culture, but it made more intriguing the behaviour I'd witnessed in the canoe that morning. Berullio was proving every bit the supportive parent, but earlier he and Antonio had shown no agitation or special concern for their son and grandson when he was alone and out of reach. Perhaps this was because they knew that sympathy at that time wouldn't have helped and might even have been counterproductive.

I asked Hermanigildo if he could feel me touching his fingertips, and he shook his head no, so I began to work. Bill held the arm up a minute to drain out as much blood as possible before I wrapped a tourniquet around it, a low-tech but highly effective technique used in even the most modern hospitals to stop blood flow completely. Without it, doing hand surgery would be like trying to fix a watch in the bottom of an inkwell.

Only now did I remove the rag I had inserted against the boy's torn artery on the beach. I washed the inside of the wound using Betadine, peroxide and rainwater. Once all the mud and jungle glop came out, I got my first good look at the damage. The cut ran a third of the way across the back of the forearm, just above the wrist and thumb. It was a neat cut; the machete must have been very sharp. I couldn't see inside, since the edges were not gaping open, so I took a long wooden Q-tip, jammed the hub of a needle over the cotton ball at the end, then bent the tip of the needle into a hook. Two of these devices made excellent retractors. I stuck one in each side of the wound, pulled them apart and gave the ends of the sticks to Bill – who was now both my scrub nurse and my OR table.

Using a forceps and scissors, I probed the wound but still couldn't see to the bottom. The cut was too deep, and while the small clearing we were in provided shelter from the blazing sun, it made for a dim operating room. The trees rose up to form a densely intertwined canopy high above, preventing the sunlight and the wind from reaching the ground. Light filtered down only in thin shafts that came and went with the movement of the leaves. Kneeling in the mud, in the soft light and still air under the vaulted ceiling, I felt the setting was more appropriate for an outdoor cathedral than an operating room.

'I need a flashlight,' I said to no one in particular. Two of our team ran down to the canoes and came back with two powerful lights. 'Try to position the beams on either side of my head and hit the skin surface perpendicularly,' I said, without taking my eyes off the wound. They directed them perfectly. The bottom of the wound was illuminated, and with some more probing I found four cut tendons and one cut nerve. There were also cuts through the radial artery and several veins, as well as a gouge in the radius bone where the force of the blow had finally been dissipated. A machete is a fearsome instrument, especially when you consider that the swing had been blunted by the resistance of water.

I estimated how long it would take to repair all the injuries. 'Can you hold steady for two hours?' I asked my human light holders and my table supporter. All three said not to worry. They were fine.

Two hours might not be enough time to repair everything, but it would be about as much time as I could hope for. I knew my three assistants would last as long as I needed them. The tourniquet was the limiting factor. There would be no blood flow to the arm during surgery and hence no fresh supply of oxygen. Limbs deprived of oxygen undergo permanent damage after about two hours, which is why applying a tourniquet can be so dangerous. After the first twenty minutes, however, the arm would become increasingly painful. With no anaesthesiologist to provide heavy intravenous sedation, my operating time would depend on my patient's pain tolerance.

Tendons are like ropes that connect muscles in the forearm to bones in the hand. When a muscle contracts, its tendon acts like a marionette string, moving the body part to which it is attached. Two of the tendons that were cut hold the wrist up; the other two hold up the thumb. Without them, the hand can't be lifted and the thumb can't be opened for grasping.

The machete had cut the radial nerve, the one supplying sensation to most of the back of the hand. It's the least important of the three hand nerves, because it's the two on the other side that provide feeling to the palm and fingertips. Nevertheless, it would be a nuisance for a seven-year-old boy to go through life not feeling anything on the back of his hand. Worse, the cut end of the radial nerve often grows into a neuroma – a ball of raw nerve endings so exquisitely painful on contact that I've had patients in New York whose arms couldn't be touched by a sleeve. I didn't want to have a patient in the jungle whose hand couldn't be brushed by a leaf.

Nerves are like wet noodles when they are cut; the ends just lie there. A tendon is like a rubber band; the end attached to the muscle snaps back and the surgeon has to go in search of it. I used a scalpel to cut farther up the arm until I could see the retracted ends, then grabbed each of them with a clamp and pulled them back. There were no frayed edges. The razor-sharp machete had made neat slices. The repairs would be relatively simple. I breathed a little easier.

From the array of packets spread out on the plastic, I selected a suture strong enough to counteract the tension in the tendons and loaded it on to my needle holder. For each tendon I sewed in and out through the centres of the two cut ends, then cinched the ends together by pulling up on the stitch and knotting it. The first one came together nicely. So did the second, third and fourth. With a much finer suture, I then made a running 'hem stitch' to smooth the edges of the repair. Tendons have to move inside the arm – they can't get hung up on the tissues they're supposed to glide past.

I was satisfied with how it looked, but I needed to see how it worked. To test the repair and also, I must admit, with an eye for the dramatic, I asked the boy to try to move his wrist and thumb. He lifted them right away with no problem. Amazonian 'oohs', 'aahs' and 'wows' arose from my audience.

Tendons need to be smooth on the outside. Nerves need to be smooth on the inside. They don't heal in the same sense that other tissues do. When a nerve is cut, its outer tube, the epineurium, has to be reconnected, but the inside part, the nerve fibres, will degenerate. For a nerve to work again, new nerve fibres have to grow through the repaired tube. The fibres look for any excuse not to grow; hitting up against a piece of suture in a repair site is often reason enough. For nerve repairs, I ordinarily use a suture so fine that it is invisible to the naked eye, but not having a microscope handy, I had to settle for a stitch I could see. Though it was finer than a human hair, it made for a relatively bulky repair. But even if it failed and the back of the boy's hand remained numb, the reconnected tube would at least contain the nerve end, making it unlikely to grow into a neuroma. All in all, the repair looked pretty good, though I had no way to tell whether it would have passed the kind of meticulous microscopic inspection I would have performed in a hospital.

Hermanigildo started to get fidgety. Until then, I had never seen a non-sedated patient tolerate a tourniquet for more than half an hour. This one had already been on well over an hour. His jungle training, adapted to the exigencies of surgery, had served him well, but he was reaching his limit.

The artery, veins and bone were still unrepaired. I planned to leave them that way. The radial artery, on the thumb side, is one of the two main conduits of blood into the hand. Either conduit is enough to provide adequate blood flow, and in fact the ulnar artery, on the pinkie side, carries even more blood than the radial. People who try to commit suicide by slashing their wrists rarely succeed because they always start on the thumb side and then lose their nerve. They would have a better chance of bleeding to death if they began on the pinkie side. Repairing the radial artery wasn't necessary, though I would have tried it if I had more time. Even if I had all the time I needed, I wouldn't repair the veins. The hand has dozens of them, yet a few are enough to maintain adequate outflow of blood. I would have repaired an important vein, but the ones cut here didn't even have names. And as for the bone, it was deeply gouged but still intact. A simple splint would do.

I quickly tied off the artery and veins to prevent heavy bleeding when the blood rushed back into them. All that remained was to close the wound. I didn't need to see inside to sew the skin, so I released the tourniquet – to Hermanigildo's immediate relief.

Bill asked someone to bring him a tennis ball from his gear

bag. He always brought tennis balls to the Amazon, he told us. They make great toys for the kids. Sure enough, our patient became totally preoccupied with the ball – rolling it up and down over his face with his good hand – while I finished sewing and applied a bandage, splint and sling. As soon as I was done, the boy got up. I tried to get up too, but quickly learned that this is not so easy when you've been on your knees in the mud for over an hour.

Though our expedition was now behind schedule, the plan remained to move the rest of our supplies today to our research camp, already set up by the crew in a clearing they had made along the lakeshore. While the others reloaded the canoes, I packed up my tackle box and ate some bananas offered to me from the crowd. Then I slipped behind one of the huts to strip down and wash off my coating of mud using rainwater from the open collecting drum. As I neared the drum, I saw a bloated, decaying rat floating inside. This was the water I had just used to clean Hermanigildo's wound. I decided I wasn't so dirty after all.

Counting out a two-week supply of antibiotic pills, I told Berullio to have his son take one a day until they were all gone. I explained how to change the bandage and wash the wound with rainwater – fresh rainwater. Sebastián translated everything for Berullio, but a little girl about ten years old was listening very intently and I got the distinct impression she would be the one taking care of Hermanigildo. His mother didn't seem to be around, or didn't bother to identify herself to me, perhaps because taking care of children is child's work.

The canoes were ready. Sebastián and I stepped inside. Berullio nodded one more time that he understood everything I had said but he hadn't asked any questions. Nor did he show any gratitude, though I was sure he was appreciative. The villagers lined the beach to watch us leave. As we pulled away I could see, on top of the mud bank, a boy with a sling on his arm chasing a tennis ball with his friends.

'What if you hadn't been around?' was the question on everyone's mind as we glided across the lake. My answer surprised them all. To a human body, I explained, a machete cut is not much different from an animal bite. The design of the hand and arm has been refined over thousands of years to absorb a blow like that and keep on going. A human with his arms at his sides or in front of him, or an animal with its front legs on the ground, is naturally positioned so that the most vital and vulnerable structures are on the inside, behind the bones, protected against an attack from the front or side.

Had the machete cut (or animal bite) destroyed every structure on the outside of the arm, the victim would be unable to lift his wrist and fingers, but he could still bend them, since the tendons and muscles that perform those functions are located on the inside. Though the victim would lose sensation on the back of his hand, he would retain it in the palm and fingers because those nerves are also located on the inside. He would bleed heavily from his veins. Because there are so many of them, however, the loss of a few is easily tolerated. There are only two main arteries, but both run deeply on the inside. Unless the blade or the attacker's teeth are strong enough to break through bone, the victim can survive with a useful claw, paw or hand, which can't open but can still close and thus is able to grip and feel a branch, a piece of food or a tool.

We reached the camp on the lake just before noon. Our Indian crew had cut some trees to make a clearing and used the wood to build shelters, tables, benches and posts to which to leash the caimans. The pile we had collected the previous night was still in the bottom of the boat. They had been watered while we were at the village and didn't seem to mind the delay. Even now, they seemed in no great hurry to get out. I stepped over them on to firm ground and then into a dry tent.

The camp was a research station for a number of different projects. The primary objective was to study the social behaviour of crocodiles. Every night, we caught them in the lake and brought them to camp, where we did all those things Antonio found so perplexing. After a while we would begin to catch the same animals over and over, giving us an idea of the total population and whether they lived individually, in pairs or in groups, and whether or not they were territorial. Meanwhile, entomologists were climbing trees to gain access to termite nests, to study not the termites but the termitophiles, tiny insects that live within termite colonies; botanists were dragging the lake with sieves to trap plankton; and biologists were doing research on plants, butterflies, snakes and any serendipitous discoveries that might catch a biologist's fancy.

My plan was to involve myself in each of the projects by being an eager assistant. My priority, of course, would always be to provide medical care whenever necessary, but if everyone stayed fairly healthy, I knew I could learn a lot about the jungle. And if not, I would learn a lot about jungle diseases.

Our camp was a pinhole in a continent-wide entanglement of plants and animals that wove themselves into a tight green fabric covering half of South America. Of the thousands of hidden lakes in this rain forest, we had set up on the shore of this one, called Zancudo Cocha, because we had heard it was infested with crocodiles. I didn't learn until after I arrived that in the local dialect Zancudo Cocha means 'Lake of the Malaria Mosquito' – a jungle menace far more deadly than crocodiles, and the single biggest reason, among many big reasons, why so few people live in the Amazon.

Malaria has killed hundreds of millions of people worldwide over the course of history. Ancient Romans believed the god Febris inflicted it, and the remnant of that idea survives to this day as the word fever. Italians gave it its name when they noted that it seemed to come from the bad air, or *mal aria*, around swamps. The cause of the disease is actually a denizen of the swamp, the anopheles mosquito. After mating, a pregnant female gets hungry for protein to make her eggs. She seeks out warm-blooded, protein-rich targets by sensing the skin odours of animals and humans as well as by homing in on the carbon dioxide in their exhaled air. After landing, she makes test holes with her needle-sharp nose until she hits a blood vessel. Then she literally spits in the wound, injecting saliva, which increases blood flow by widening the vessel and preventing clot formation. The female anopheles has a distinctive style of blood-sucking, lifting her tail high in the air for the two to three minutes it takes to fill her tank. She won't need a refill unless she gets pregnant again, and until then she'll go back to feeding on plants – a much safer way for her to eat, since she runs no risk of getting swatted.

The mosquito wreaks havoc not because of the blood she sucks (only about one drop) but because of the dirty needle she uses to withdraw it. The mosquito's nose acts as a conduit for parasites that flow into the bloodstream along with her saliva. They find a home in the animal or human liver, where they thrive and multiply for the next couple of weeks or months, the incubation period, before breaking out into the bloodstream in full force to ride in red blood cells and destroy them, causing fever and chills. The debris flows into various organs, clogging them up and sometimes causing fatal damage. Even when a malaria victim survives an outbreak, a nest of parasites always remains in the liver. Untreated victims can look forward to periodic outbreaks of the disease for their entire lives.

The appropriateness of the lake's name was immediately obvious. Of the uncountable number of insects that bit me during my time in the Amazon, however, never did I see a single one lift its tail in the air. In any case, we were all taking chloroquine, a synthetic pill that protects against most forms of malaria. There is no truly effective protection against the bites, however. Bugs in the Amazon are tough, and they laughed at our lotion or spray repellent. Some of them, judging from their reactions, seemed to need it to live.

Despite the bugs, camp life was pretty comfortable. Every night a team of crocodile catchers roamed the lake. I got used to having breakfast each morning alongside a fresh batch of black caimans leashed to poles, waiting to be weighed, measured and tagged before being dropped back where they came from. Besides catching them, Antonio proved indispensable to the recording procedure, as I saw one morning when our researchers weren't sure whether one of the crocodiles was male or female. Antonio did a rectal exam on the protesting animal. 'Macho', he announced.

Antonio was indispensable in other ways as well. Along with being the leader of our crew, he was highly respected among the Cofan Indians as a repository of rain forest knowledge. Since his primary involvement was in night-time crocodile patrols, he was often free during the day, as was I, when no one needed my attention and I wasn't assisting one of the biologists. He had never thanked me for sewing up his grandson, but he seemed to pay me special attention.

Except for the usual jungle complaints – dehydration, diarrhoea, itching and fungus, our team remained pretty healthy. Word of my surgery on Hermanigildo had spread, however, and before long I was receiving office visits from villagers arriving via back trails or pulling up in canoes. One boat approached bearing two girls and a little boy in the front and a woman paddling in the back while holding a baby on her shoulder. She guided the dugout canoe on to the shore as smoothly as if she were parking the family car. Everyone got out, leaving a rifle and a baby bottle on the seat. My patient was the little boy. His mother had brought him because for weeks he had been crying, not eating and having diarrhoea. Sebastián asked a few questions for me, and I noticed that whenever the mother hesitated, the older sister jumped in and answered. Like Hermanigildo's sister, she seemed to be in charge of the baby; it was apparently just one of the many tasks routinely delegated to young girls. My examining table consisted of a row of heliconia leaves (large flat ovals, each a metre long) laid side by side right on the ground, which since it had been cleared was getting muddier after each rain. The boy looked uncomfortable; his stomach was a little bloated and his liver felt enlarged. He might have dysentery or, worse, he might have schisto – yet another reason the jungle is an extreme environment.

Schistosomiasis is called 'snail fever' because it is a worm disease carried by snails that lie in muddy river bottoms. There they breed the parasite, releasing over a thousand a day into the water. The worms go in search of soft, thin skin they can bore into, such as the webs between the toes of little boys wading in rivers. Like most parasites, they make their way toward the liver. All food absorbed by the intestines is processed by the liver, which is the most nutrient-rich place in the body and therefore the most desirable neighbourhood for a parasite. Male and female worms meet in blood vessels near the liver, forming a union that can last ten years and produce several hundred eggs per day. The offspring enter the liver, then, flowing into the bowels, they cause diarrhoea, which expels them from the body and, if the toilet is the river, back to the snail. Successful parasites merely weaken their hosts, keeping them sick for years; killing the host would ruin the parasite's lifestyle.

I treated the boy for dysentery. I couldn't have treated him for schisto even if I'd been sure he had it. I didn't have the medicines, which in any case are dangerous to dispense outside a hospital; and even if he were cured, he'd undoubtedly go right back to the same river and be reinfected.

As the family was leaving, the little boy – who until then had not shown much vitality – asked if he could have a tennis ball. The word had gotten out. We gave him two. Holding one in each hand, he stepped easily into the canoe, satisfied with his visit. Maybe he wasn't so sick after all.

Another canoe arrived one evening with an emergency trauma. An elderly appearing woman (though it was hard for me to judge age from the faces of these jungle-worn people) had been knocked in the head while attempting to tackle a pig she intended to prepare for dinner. She had been stunned and dizzy at first, and now had a headache. I did a complete neurological exam and then wanted to check her eyes. As I lifted my ophthalmoscope out of the tackle box, I found a cockroach running around inside. I unscrewed the handle of the scope, then used the bottom end to crush the roach. Neither my patient nor my audience of crew members thought anything of this; it was all quite natural to them. I imagined what the reaction would have been had this happened at home. My patient passed her exam easily. This was unsurprising, since she had canoed over by herself. I told her there wasn't anything wrong with her. 'I know,' she replied. 'I just came for something for my headache.'

Antonio often watched me work. Once, after I had treated a routine problem, he came to me with a round, spiny fruit called an achiote. He broke it in half; the inside was filled with small, bright-red seeds. At first I thought he was offering me a snack, but then he pressed his thumb into the middle of one of the halves, squashing all the seeds, creating a cup with a red paste inside. He put some on his fingers and painted my face, drawing an upside-down T on my forehead, a line under each eye and a line across my chin. The dye went on as easily as if he were using a grease pencil. 'This makes you a witch doctor,' he said. The Indians watching him laughed. Antonio laughed as well, but I could sense his seriousness; his laugh, I suspected, was to show the younger generation that he understood he was hopelessly old-fashioned. I told him I was proud to wear the markings.

Antonio said that witch doctors used face paint for its psychological effect on their patients and that this was a big part of their medicines' efficacy. I told him it was a big part of ours too. Our doctors don't wear face paint, but they do wear white coats, to keep that same separation between themselves and the people they treat, and they often use a big medical word when a simple one would do, to give an impenetrable, mystical quality to the knowledge they possess. Maintaining that aura of authority is essential, so that a sick person can believe his doctor has the power to make him better. Much as I would have liked wearing my witch doctor paint every time I saw a patient in the jungle, it would have looked silly in a camp full of scientists, amid natives who wore clothes that looked like leftovers from a Kmart closeout sale. Though they had practical value, these garments were exotic artefacts produced in mysterious ways by a culture that obviously controlled powers far greater than those controlled by the local medicine man. Their proliferation eroded respect for traditional customs and polluted the jungle.

In spite of our Western costumes and our comfortable camp, however, we never forgot we were in the jungle. The daily rains and steam-bath humidity turned our campground into a slippery, muddy ooze. It was impossible to stay either clean or dry; one by one we succumbed to the allure of the Indians' bath boat.

Every afternoon the Indians took a canoe out to the centre of the lake, soaped themselves up and, to our astonishment, dove into the water to rinse off. We had caught, and released, more than enough crocodiles to know what was in the water with them. Moreover we had been catching and eating piranha and weren't anxious to give them an opportunity to get even. Yet the natives swam about unconcerned. They live here, we reasoned; they must know what they're doing. And anyway, we were so hot, dirty and uncomfortable that suicide in the lake was starting to look like an option. Their explanations made sense – or perhaps we were ready to believe anything. They told us that the crocodiles, dangerous as they are, spend the day in the shallow grasses along the shore, only venturing to the centre of the lake at night. As for the piranha, they are meat-eaters with razor-sharp teeth that intermesh to close like a bear trap. They're small but travel in groups of thousands. Piranha sense blood, and when they're hungry it drives them into a frenzy. They can strip a mammal to a skeleton in minutes. (So far, their words were not too reassuring.) However, they added, during the rainy season the lake expanded, and fruit trees that were on land during the dry season were now standing in water, dropping their fruits into the lake where the fish can reach them. Piranhas much prefer fruit to flesh, so as long as they remained finicky eaters, there was nothing to worry about.

Several of us hitched a ride on the next bath boat, preparing to dive in before we came to our senses. First the Indians pounded the lake bottom around the canoe with long poles. This was to chase away any stingrays that might be lurking in the mud, we learned. A stingray has a tail longer than its body, and at the end of it sits a rigid stinger lined with barbs and coated with venom. The tail acts as a muscular whip, propelling the harpoonlike stinger deep into whatever disturbs it – such as, for example, the foot of a doctor desperate for a bath. The injected toxin causes severe pain and significant tissue damage. It can even be lethal, though rarely. I had heard that holding a lit cigarette near the wound was very effective in relieving symptoms because the venom is heat-sensitive, but no one on the boat smoked and, all things considered, I thought it wiser simply not to step on the bottom.

The natives didn't seem concerned about still another inhabitant of these waters – the electric eel. I was determined to swim, so I decided a visitor could ignore it too. The eel can grow to 3 metres in length and is mostly made of muscle. Every animal's muscles produce some electricity when they fire, but the eel native to the Amazon basin has modified its muscles to produce a lot more. Plus, they're hooked in series so the voltage is additive. When it's annoyed, the eel can produce a 500-volt electric field around itself. The Indians say they feel a tingle when one of them swims close by. You don't have to touch it to be electrocuted – you just have to violate its personal space.

We were all ready for a carefree swim. Just as we were about to dive in, Bill said, 'Okay, no skinny dipping.' Who asked him? I thought. Although our canoe was mixed, I seriously doubted that anyone would have minded, or even noticed, considering how many other things there were in this lake to occupy our minds.

I hit the water and it felt great. Amazon water is very soft because the soil through which it runs is mineral-poor. The swim was so refreshing, I couldn't help but relax. I didn't want to get out of the water – until something suddenly grabbed my leg. I recoiled, ready to fight for my life. My first thought was 'Can I reach the boat before the crocodile pulls me under?' My second thought was 'These Indians are real jokers,' as I saw one surface in front of me, laughing.

I got back on the boat anyway. Bill was there already. He said he didn't know how any of us would get our underwear dry. 'If you hadn't said, "No skinny dipping,"' I pointed out, 'we wouldn't have the problem.'

'It's not because I'm a prude,' he replied. 'It's because there's candiru in this lake.'

Just the thought provoked a sharp pain in my groin. The candiru is a kind of catfish, about the size and shape of a toothpick. Attracted to salt, such as that contained in the urine within a human bladder, it is small enough to swim through a male or female's genital opening and get lodged in the urethral tube. The fish's stiff pectoral fins angle backward; there is no way to pull it out. It has to be removed surgically. So a little modesty can be good preventive medicine.

These medical facts I was already familiar with, but Berullio volunteered some additional information. He swore that the fish's attraction to salt was so strong that he had once seen one swim up the urine stream and enter the penis of a man peeing in the water. I dismissed the story as an example of an Amazonian Indian having fun with a credulous American doctor. Yet competition for survival within this lake was desperately fierce. Only highly developed adaptations kept players in the game. The anopheles mosquito, black caiman, schisto worm, stingray and electric eel have each found their own unique way to succeed. So has the candiru, whether or not it can swim up a urine stream.

Survival in the Amazon requires everything from wearing underwear to learning how to handle poisonous snakes. Coming back from the camp latrine one evening, I was about to step over a bent branch hanging low over the trail when suddenly it moved. I caught my balance, stopped in my tracks and felt my heart palpitate. My headlight illuminated a deadly black coral snake. It was coiled around some leaves, holding its head out and straight up in the middle of the trail about 30 cm off the ground. It had a tiny ugly white head and a body of alternating bands of black, orange and yellow that stood out against the green leaves before disappearing in the dark underbrush. The sharply contrasting colours, lit up by my headlight, made the snake hideously attractive – almost transfixing – while it rhythmically bobbed its head up and down and flicked its tongue. I had caught its attention, and it had definitely caught mine. But there was no need for us to scare each other. I stepped back a few paces, made a very careful semicircle and then hurried down the trail to camp.

The snake was still there when I returned a few minutes later with Antonio and several others. Antonio poked at it with a long branch, got it to wriggle on to the bushy end of it and carried it back to camp, laying it on a log next to the dining table for everyone to look at. Then he killed it. With one whack from his machete he separated the head from the body. I jumped at the unexpected violence and looked questioningly at Antonio.

'No way to tell if it was poisonous, and too close to camp,' he explained.

Pushing the back of the snake's head against the log with one hand, he poked at the mouth with the end of a banana. Though the head was completely detached, it bit down viciously into the fruit. The survival instinct lived longer than the snake had.

Antonio was not one to let a good snake go to waste. He turned the body over, slit the belly open and scraped out some highly unappetizing ingredients, which he collected.

'Much easier to skin a snake when it's fresh,' he said.

I wondered how anyone could eat that goop, and Antonio caught my look of revulsion.

'Not for eating,' he said. 'For bait.'

I was reassured, but then he added, 'Too small to cook.'

Just the head was left on the log now, still clamped on to its prey. With some difficulty, Antonio extracted the banana and showed us the bite marks. There were two rows of dots that nearly formed an oval, and at one end, just outside the oval, were two large holes – the teeth and fang marks of a poisonous snake. I would have liked to examine the head further, but Antonio said it was still dangerous and threw it into the jungle. I thought he meant it could still bite, but he said, 'Snake heads can cast bad spells.' He didn't seem to be joking.

After dinner, sitting around our campfire, Sebastián told me the terrifying story of an encounter Antonio had had with a snake many years earlier while working for some geologists from an oil company. His job was to guide and protect the vulnerable strangers who, like us, had no chance of surviving on their own. They were carrying out the kind of work that was destroying his homeland, but Antonio saw no conflict. He had learned from his ancestors that the environment always recovered, and had seen this for himself his entire life. He didn't realize that this time he was opening the door to a civilization with far more destructive force than anything he could imagine.

One cool night in that tiny jungle-encircled oil camp, Antonio was asleep on the floor of his tent, wearing loose clothes and lying on his back between two blankets. He was awakened by some wriggling around his ankle followed by the unmistakable sensation of a snake slithering up his leg inside his pants. He felt its full weight as it coiled up to rest on his stomach. Reflexively, Antonio pulled the blanket off, but the snake was under his T-shirt. Snakes are cold-blooded, meaning that their body temperature depends on their surroundings, and on cool nights pit vipers have been known to enter tents in search of warmth. A pit beneath each eye acts as a long-range infrared heat sensor, easily able to detect the warmth coming out of a tent. With its ability to distinguish temperature variations as small as 1°C between its right and left pit, the snake can target any heat source and home in on its warmest part; in this case, Antonio's stomach, just below his heart. The snake had only been seeking a comfortable place to rest and had no intention of biting him. But snakes, though deaf, are exquisitely sensitive to motion or vibration. Antonio knew that if he changed position or called out, the snake would interpret it as an attack.

If it were going to strike, it would lift its head up, as if cocking its upper body, before plunging its fangs forward and down faster than the eye can see.

The snake remained motionless. For hours, Antonio lay on his back watching the bulge covered by his T-shirt, afraid to either move or yell for help – afraid to see the T-shirt start to tent up from below. Just after sunrise, the tent and the snake began to heat up. Soon the snake became uncomfortable and started moving around. At last Antonio felt it slide slowly past his thigh, down his other leg, and out over his foot. He waited a few more seconds before he dared to get up, and when he did, the snake was gone. He never saw it.

Later I asked Antonio what kind of snake he thought it was. He believes it was a small Cascabel – a deadly poisonous pit viper. He said it was about a metre long and very heavy. Its scales felt rough against his skin, and it was black and grey. I asked him how he knew it was black and grey if he never saw it.

'I see it in my dreams,' he replied.

I wondered if he was thinking about that when he told me that snake heads cast bad spells. For me, and for most people, simply looking at a snake is enough to elicit a mixture of chills, fear and revulsion – a dramatic, but not altogether inappropriate, survival response, considering that poisonous snakes kill over one hundred thousand people per year, roughly half of them in India. No wonder some Indian religions worship snakes as gods to be feared.

Poisonous snakes are highly evolved killing machines. Their venom is a special saliva formed in glands behind their jaws. During the bite, the glands contract, squeezing the venom forward into the fangs, which are hollow teeth with holes at the pointy ends. The system is very much like an injection with a hypodermic needle.

Venom is a complex potion, not easy for a snake to produce. Therefore the snake has to use it sparingly, and only for very good reason. While the snake will always inject venom into prey, when biting defensively it may or may not envenomate its assumed attacker. It first quickly assesses the risk in order to decide whether using up venom is necessary. When humans are envenomated, it is almost always because the snake felt mortally threatened, from being stepped on or poked at. Snakes do not want to waste their venom on humans, which are not prey. Because a snake's teeth point backward, they are good only for gripping; snakes can't tear apart or chew their prey, and humans are too big to swallow whole.

Venom is a mixture of poisons and of enzymes, substances that dissolve body tissues. Each kind of snake makes its own variety; the effect of venom inside the body therefore depends on its individual recipe. The coral snake I met near the latrine uses mostly nerve poison, while the snake that slept with Antonio prefers to poison the blood. Both types are quick and deadly because they attack the body at its weakest points – the fragile chemical reactions critical to the dynamics of breathing, heartbeat and blood clotting.

Muscle contractions make hearts beat and lungs breathe. Muscles are powerful machines that can sustain a lot of damage and keep on going. But like a lot of machines, they will stop cold when unplugged. They depend on the constant supply of electricity that reaches them through the wiring system of the nerves. Small spaces exist between each consecutive nerve and between the nerve and the muscle. When the current comes to the end of a nerve, it faces a gap it has to jump if it is to continue. The nerve end releases a chemical called a neurotransmitter that floats across the gap and sparks the next nerve. The last nerve before the muscle ends in a group of tendrils, like a multipronged plug, that align with but don't quite connect to a corresponding socket in the muscle. Another neurotransmitter has to float a connection between the two. It is precisely this last delicate step that the coral snake's venom blocks. The venom mimics the neurotransmitter and fills the gap, but it does not conduct the signal. The transmission of electrical impulses is interrupted; muscles fire irregularly, then weaken, and finally become paralysed. The victim twitches, collapses and loses the ability to speak. Within a few hours the paralysis overtakes the pulmonary and cardiac muscles. The lungs stop breathing and the heart stops beating.

The pit viper uses a different but equally lethal technique. Its poison weakens the walls of blood vessels, allowing blood cells and fluid to leak into the surrounding tissues. Whenever a blood vessel is breached, the body's natural defence is to set off a cascade of reactions that convert free-floating raw materials into a sticky mesh that forms a clot to seal the hole. That process, known as coagulation, is why a cut stops bleeding. Pit viper venom sabotages the intricate sealing process with an ingredient that mimics one of the crucial raw materials and gets itself incorporated at an early step. It is much weaker than the real thing, so the final clot ultimately gives way. The mesh cannot form a seal strong enough to stop the bleeding. It is a fatal weakness: the blood goes everywhere but where it belongs, filling the lungs and emptying the heart.

Because venoms are thick, they contain an additive, hyaluronidase, that helps diffuse their poisons. Doctors routinely add this dissolving enzyme to injected medicines to increase the rate of absorption. Snakes use it to speed up the spread of their venom. Snakes also inject another enzyme that digests muscle. It causes the severe tissue destruction often seen at the site of a bite, but its real purpose is to give the snake, which cannot chew, a head start on its anticipated meal.

Treating someone who has been on a snake's menu is not easy. If he has been envenomated, he will probably die. Venom is not a single poison; one injection is similar to a multidrug overdose. The catastrophic breakdown of many organ systems that rapidly follows the poisoning is hard to reverse, even in the most sophisticated hospital. In the jungle it would be impossible. But less than half of all bites from poisonous snakes actually inject venom. Fang marks are sometimes hard to see, particularly because they are often obscured by swelling. Tingling, facial numbness, palpitations and difficulty breathing can be early indicators of envenomation, but given the near hysteria aroused by snakebites, those signs can be present in anyone.

If I were treating a snakebite, my first job would be to calm the patient down – not just for the psychological benefit but to slow the spread of the venom. At Zancudo Cocha I had an extractor, a plunger type device that if applied to the bite wound within the first few minutes might draw out some of the venom. It might also be a waste of time. No one knows for sure how effective it is. I would not try sucking the venom out by mouth, though not because I would be afraid of swallowing poison. As potent as they are, these poisons are easily broken up by stomach enzymes. Snake venom only works when it enters the bloodstream directly. Anyway, venom is so thick that sucking it out by mouth would not work. Even worse would be making a cut over the bite and trying to squeeze the venom out. That would expose more blood vessels to the venom and probably drive it in deeper under pressure. I could slow the poison by applying a wide bandage over the whole limb. With just enough pressure, I might be able to block the veins from carrying contaminated blood to the heart while still allowing inflow from the deeper arteries, which would carry fresh blood into the limb to keep it alive.

I didn't bring any antivenin on this expedition but not because I forgot. The most effective antivenins are specific for each species of snake, and there are over one hundred varieties of poisonous snake in the Upper Amazon Basin. General antivenins are not as effective. Many bottles are required, the injections themselves are dangerous to give outside a hospital and, worst of all, they have to be kept refrigerated. Antivenins are antibodies, natural body chemicals that patrol the bloodstream and deactivate antigens – foreign chemicals, such as snake venom, that find their way inside the body and cause havoc. Each antibody is specific to one kind of antigen and only manufactured in quantity once the body has been exposed to it. When enough is manufactured, it can counteract the antigen and stop the problem. This is how we 'get over' a cold. But snake venom, like many deadly diseases, doesn't allow enough time for the antibody response. The solution is to have the antibodies ready to go – premade by injecting sublethal doses of venom into animals and then collecting the antibodies they develop. This is how vaccines work. But antivenins have a lot of impurities and can create their own deadly side effects. After consulting with experts at the Bronx Zoo, I decided they would be too impractical for this expedition. People on expeditions understand they assume risks when they enter an extreme environment. Practical decisions I make may have fatal consequences. The responsibility would be mine.

There would be nothing more I could do if a snake sank its fangs into a member of my team. Evacuation was impossible. We were surrounded by thousands of kilometres of trees and plants . . . but maybe one of them held a cure for snakebites. I asked Antonio what he would have to offer if one of us were bitten. He said he once treated somebody who had been bitten by a bushmaster – probably the deadliest snake in the Amazon. He made a poultice from the inside of a tree bark, added a few plants, then boiled it until it had turned into a syrup. He gave the victim one dose every day. The plan was to give a total of five doses, but on the third day the patient died.

'Yeah,' I mused. 'I don't know why that happens. Sometimes you follow the treatment protocol exactly, and the patient still doesn't get better.'

The laws of nature are sometimes simply overwhelming, and doctors are powerless to intervene. Bodies have limits. Though the people who live in the Amazon have adapted to their surroundings, they have done so mostly on brainpower. As evolving organisms, they are still works in progress, having not yet acquired adequate physical protection against their most extreme hazards. They remain as vulnerable to snakebites in their environment as we are to automobile accidents in ours. Where one mistake can often prove fatal, survival results not from experience but from being taught, whether the lesson is how to spot a snake or how to cross a street.

Out of the forest one day appeared an Indian who had travelled three days with his family to see me. He had a severe limp, and it had obviously been a difficult trip. He told me that his leg had gotten weak after a childhood sickness, then stopped growing. His left leg was badly atrophied and several inches shorter than his right. The foot was angled down and rigid, touching the ground only at the toes. He hadn't been this far from his village for years, but he had come because he had heard about my surgery on Berullio's son's arm and thought that perhaps I could help him as well.

With one look I knew that I could not. His childhood disease was probably polio, and his deformity, after all these years, was permanent. The best treatment would be amputation, followed by crutches and a prosthesis, but that would be absurd here. He would have to leave the Amazon. With a wife and three small children, he had obviously adapted well to his condition and had no intention of going anywhere else. There was nothing I could do for him.

Well, almost nothing. He said his entire family was with him and asked if I would take a picture of them. A family photo. I was happy to oblige. Father and mother in back, the three children in front, all of them taking a serious Victorian pose – I used a Polaroid camera and handed them the snapshot as it came out. Without my telling him, the father knew to hold it at the edges until it was fully developed. I got the distinct feeling that this was not the first Polaroid photo he had, nor was I the first doctor he had seen.

My medical practise was steady, but not every problem was serious. Though all the biologists wore boots, and most of the Indians did not, we were the ones who developed foot problems. To the endless amusement of the native crew, I treated blisters and foot fungus caused by boots that couldn't be kept dry on the inside. They teased us by pointing to their bare feet, thickly calloused from a lifetime of walking in the jungle. Tenderfoots like us always considered calluses a nuisance, but here they were nothing less than an adaptation for survival.

One member of our team made the mistake of lighting a fire under a small tree while out in the field. It smoked out some stinging caterpillars, which dropped from a low branch. One fell harmlessly in her drinking cup but a few others landed on her arm and neck. The stings themselves weren't so bad, but now she was getting itchy. Native to the Upper Amazon Basin are iridescent caterpillars with venomous spines that can deliver fatal stab wounds, but the ones that attacked this biologist were the multi-coloured kind and only caused a simple irritation from the sticky hairs still adhering to her skin. The treatment was a strip of duct tape that I applied as if I were removing lint from a jacket.

For one complaint I found it hard at first to maintain a serious medical demeanour. An expedition member came to me, pointing to his buttocks. 'There's something moving around back there,' he told me, unnerved and embarrassed. He pulled his pants down and bent over, and I undertook a flashlight search. Each buttock featured several white pimples with black plugs in the centres. And each of the black plugs was wiggling. It was an infestation of tungas, small black fleas that hang out on the ground in areas of cleared brush such as that around our latrine. When offered a target as tempting as bare buttocks resting just above the ground for several minutes, they literally jump at the chance, tunnelling into the skin headfirst until only their tails are left protruding. This convenient arrangement allows them to suck blood and defecate without changing position. They grow to pea size, and when ready to lay eggs, they wriggle around, causing intense itching. The scratching that follows helps release and disperse the eggs. One of nature's more lowly animals, the tunga has nevertheless developed a very clever survival strategy.

I applied a cotton ball soaked in alcohol to each protruding flea tail and left it in place for several minutes until I was sure the flea was dead. Then, with a scalpel, I made a slit across the hole at the centre of each pimple and folded back the skin on either side to expose the body containing the egg sac. Using tweezers, I pulled it out carefully; rupturing the egg sac could lead to a severe infection. Once the body was removed, the head became visible, embedded in the tissues firmly enough to have snapped off when I pulled on the body. With no fragile egg sac to worry about, I could pull much harder to extract the stubborn head.

Naturally I was prepared to maintain medical confidentiality regarding the identity of my patient, but soon realized that being part of the life cycle of a tunga was a fascinating experience for a research biologist. The news produced a stimulating academic discussion, quite a bit of anxiety – and some odd behaviour around the latrine. Besides spraying the area mercilessly, we all meticulously applied bug lotion to those body parts we intended to expose, and at night extinguished our headlights to black out the target once we settled in place. Two people even developed a buddy system, accompanying each other to the latrine to take turns fanning away fanny fleas.

The tungas were a reminder (as if the crocodiles, snakes and piranha weren't enough) that we were in hostile surroundings. Our camp was a fort, but it provided us with only partial protection. We were subject to all the natural laws of this environment – including the weather. Though I could no longer remember what life was like without rain, Antonio told me that the season had been unusually dry and water levels were dropping. Since our remote lake was only accessible by canoe during high water, it was becoming increasingly difficult to get supplies in to us. Though I didn't sense any threat to my survival, I was nevertheless feeling increasingly isolated from civilization. The imperatives of the jungle were closing in.

Survival is a powerful motivator for change. Animals survive over time by adapting their bodies and their behaviour to changing environmental conditions. Human beings adapt as well, but they haven't progressed nearly as far as animals – except in brain development. In fact, so powerful and so nimble have human brains become that they hamper physical adaptation, since they have the capacity to adjust to environmental pressure by mental agility. The pressure of natural selection to bring about physical adaptation is deflected by conscious behaviour modifications. Though my situation at Zancudo Cocha was not immediately life-threatening, the Amazon was becoming increasingly intimidating, and that was enough to alter behaviour that until then I had thought ran deep. Learned responses fall away quickly when they are removed from the surroundings that formed them. I was about to take my first conscious step toward adaptation. It took place over a bowl of spaghetti.

In the jungle, a meal belongs to whichever species can eat it first. The spaghetti served to me one evening was covered within seconds by a swarm of tiny black insects. I would have thrown the food away except that I knew there would be no second serving. Faced with the choice of eating it or going hungry, I picked up each strand of the spaghetti, drew it through my thumb and index finger to slide off most of the bugs, then ate the strands one by one, including whatever bugs were left on them. In a small way, I was adapting to a new environment, overcoming years of cultural conditioning that seemed supremely irrelevant here. I knew that once back in New York, I would quickly return to being a fussy eater.

My mind was adapting to the jungle in subconscious ways as well. One particularly hot night I was finding it impossible to sleep in the heavy humid air inside my tent. Every breath was laboured, but I was too tired to get out of my sleeping bag to zip open the tent flap for ventilation. My tent mate, luckily, was less tired, and in the pitch dark I heard him get out of his sleeping bag and unzip the zipper. To my great relief I soon felt much cooler and drifted off to sleep. I awoke in the morning feeling refreshed, but puzzled that the tent flap was closed. My tent mate told me that the zipper noise I had heard had been him taking something from his backpack. He had never opened the tent flap.

The sound of the zipper had been enough to cool me down only because I had so completely misinterpreted it. Being convinced the flap was open had brought about the same response as actually opening it. When the mind believes something strongly enough, it can will the body to make it so. My survival instinct was beginning to stir. I was gaining command over my body, forcing on it the changes in behaviour that I needed to adapt to my surroundings. Though my hardships were minor, they were enough to activate my will because I was apprehensive in this extreme environment. I had already seen the power of the human will in a seven-year-old boy with a deeply damaged hand. His injury should have generated overwhelming pain and panic, yet he remained calm and purposeful. He didn't need to feel pain to know his situation was serious, and by suppressing the pain he had vastly increased his ability to help himself.

Several times I asked Berullio how his son was doing and told him to have the boy come to camp for a post-op follow-up. Berullio always said his son was okay, and though he would occasionally leave camp and return the next day, he never brought the boy back with him. Antonio, the boy's grandfather, explained why. Berullio didn't live in the same settlement with his children. Years before, he and his wife, Antonio's daughter, had hired an outsider to help them with their farm. The farmhand ended up taking more care of the wife than the crops, and she ran off with him to the big coastal city of Guayaquil. Berullio wasn't much interested in raising his children by himself. He moved in with a woman from another village, leaving the children to be raised by other relatives. This wasn't the first time I had seen family problems interfere with a patient's care, but I had never expected to find it in the Amazon.

I suggested I make a house call. In addition to being good medical practise, it would give me a chance to see more of the land and meet more of the people. Antonio didn't think the trip necessary but readily agreed to take me. He was willing to unlock a door for me because I was the *doctoro*. The trip would take a few hours. With the water level dropping, it would be difficult to go by canoe. We would have to cross the lake and then take an overland 'short cut' trail.

At sunup, I packed a few medical supplies, put sulphur

powder in my underwear and socks to keep the ticks out and applied bug lotion everywhere, though it only seems to encourage them. Antonio guided our canoe across the lake, then worked his way up a hidden channel of flooded land with trees standing in the water. The channel was green, the trees were green, even the air was green. The surroundings got narrower and gloomier as the canopy enclosed us. We scraped against the trunks of partially submerged trees whose branches were so close to the water that we had to lie flat on our backs as the canoe glided under them. Looking straight up as we slid under a branch that barely cleared the canoe, an anaconda came into view directly above my eyes. It was coiled up in a notch, sleeping peacefully. Slowly it passed over my head, giving me ample time to imagine its reaction were it to be shaken loose from its tree and suddenly awaken in the bottom of a canoe.

An anaconda is not poisonous. It kills by biting down, gripping its prey with sharp, backward-pointing teeth, then coiling itself around the victim's chest. Each time the prey exhales, the snake tightens its grip until the animal suffocates. It then swallows the victim whole, headfirst. Anacondas can be as wide around as a tree trunk and can swallow an animal larger than itself. To open wide it will unhinge its jaws by relaxing the muscles that keep them locked together. For a really big meal, the body can enlarge as well, by expanding the ribs outward; snakes have no breastbone to hold the ribs in place. Antonio told me that he had once seen a 9-metre anaconda eat a 4-metre caiman. Even after the snake swallowed it, he could see the bulge of the crocodile within the snake's distended body. Not one to miss an opportunity, Antonio killed the snake and went home that day with two prizes.

'So,' I declared, my imagination in overdrive, 'that snake we saw could have swallowed the whole canoe, bow first, with us inside.'

Antonio shook his head no, ever patient with his pupil. 'Adult anacondas don't sleep in trees. That snake was just a baby.'

And so, I had just met another of Antonio's neighbours, one that survives by crushing its prey. Anacondas either developed powerful constrictor muscles because they couldn't make venom, or else once they had the muscles they no longer needed to make it. Everyone living here seems to specialize in one survival technique.

The channel ended in a mix of weeds and grass. As Antonio was tying up the canoe, he cut his finger on a razor-sharp blade of sawgrass – a small cut for which I handed him a Band-Aid. He responded by handing me some red achiote paste. I started to apply it to my face, though it seemed a little silly to become a medicine man just because I had given him a Band-Aid. He laughed and pointed to my ankles. Achiote paste has a more prosaic use than as war paint, and before we entered the forest, he wanted me to smear it around my ankles. It would work better than my sulphur powder to prevent ticks from crawling up inside my pants legs. There was still some left when I finished. Antonio smiled and pointed to my face. All the same, I put the rest in my pocket.

Hiking in the jungle always gives me the feeling that I am inside some giant organism. Trees encircle me and plants cover every spot on the ground, but there's no sense of order: branches, vines, leaves and stems mix chaotically, surrounding me on all sides and overhead. Tree trunks rise up, opening like giant umbrellas, their branches tightly interlocking before disappearing into the canopy above.

The morning air was cool and moist, the light dim but attentive to every variation of green. Occasional glints of sunlight speckled the ground, forming kaleidoscopic patterns that rolled over us as the branches swayed overhead. If there was a trail here, it was invisible to my eyes. Though Antonio had come through here a week ago, the forest had already taken back the trail.

He set about carving it out once more. Leading the way with his machete, he cut a path just wide enough for us to walk single file and just high enough for him to pass under. Antonio was a foot shorter than I was and did most of his cutting in a semi-crouched position, creating a low tunnel through which I advanced with bent knees and stooped shoulders. The machete never stopped moving. Antonio was equally adept at forehands and backhands, though I think some of his backswings and follow-throughs brought the blade a little closer to my face than he realized. That thing was really sharp. It sliced through loose dangling vines, dropping them straight to the ground. I admired Antonio's skill. There was no wasted motion. His technique reminded me of the chief hand surgeon at Columbia Presbyterian Hospital in New York, where I had done my fellowship. Though he never seemed to move quickly while operating, all his actions were so precisely placed and maximally efficient that he could perform an operation faster than anyone else. Both he and Antonio had applied their similar talents to learning skills appropriate for their environment – the master surgeon in New York, the consummate trailblazer in the Amazon.

The seemingly endless and impenetrable forest gradually relented, giving way to a foul-smelling swamp. The hard part wasn't crossing the water; it was crossing the muck to get to the water. Mud sucked at my boots; I had to pull hard at every step to extract them. Preparing to jump over one muck-lined rivulet, I pushed down hard with one foot and lifted the other. I heard a loud sucking noise as one leg sank in over the boot top and the other came out of the boot entirely. I managed to get it back in, but all my struggling served only to embed me deeper.

This wasn't quicksand, but it was close, and as with quicksand, the danger is not so much of drowning as of getting irremediably stuck. If you don't struggle, your head won't go under, but you can sink in up to your chest, and with no way to defend yourself, and your head at ground level, you are easy prey for anything that crawls or slithers by. Intentionally burying someone this way is a custom some jungle tribes once reserved for their worst enemies. I was beginning to understand what it felt like.

Antonio looked back at me and smiled like a tolerant uncle, then began casually chopping down a small tree. After he had trimmed off the branches, he handed it to me to use to pry myself out, which I managed to do. The deeper slime-covered pools all seemed to have fallen trees lying across them. It was no accident, I realized. This was, in fact, a trail. Crossing log bridges on mud-slicked boots is pretty tricky, but I overcame the repeated challenges by concentrating on my feet and the placement of my hiking pole, steadfastly refusing to consider what awaited me if I slipped into the unfathomable slime on either side.

Gradually the land rose and a path became evident even to me. The trees closed in again and I felt confined, but we were in a corridor now, moving more rapidly than before. Antonio skimmed over the trail with a smooth, shuffling motion as if feeling the ground with his feet. He never looked down; his eyes roamed the canopy continuously. He was alert to any sight or sound that might be interesting to me or valuable to him. At times he would stop his machete in mid-arc and point to something that would take me a long time to find – a family of howler monkeys or a marmoset; hummingbirds, macaws and toucans; translucent butterflies and delicate purple flowers. I was seeing the jungle through Antonio's eyes.

And hearing it through his ears. A sharp crack sounded as a large branch broke off an unseen tree, followed by a muffled woomph as it landed on the roof of the canopy above us. Parrots screamed in flight. Bushes rustled from unseen animals. All the typical jungle noises I remembered from old Tarzan movies were now being brought to me with startling immediacy in three dimensions.

I was experiencing the jungle through Antonio's senses but not through his mind. I was a jungle tourist. Antonio showed me the riches around us, but I saw the plants and animals as natural wonders. For him they were resources or dangers, or both. Even had I been suddenly endowed with Antonio's heightened senses, I'd be clueless about how to use them; and human senses, even when brought to their fullest potential, are still meagre compared to those of the animals with which the humans compete. The far more critical factor in adaptation to the jungle is the vast storehouse of knowledge the brain is capable of holding. I would need a lifetime in the jungle to acquire that knowledge and to develop the skills to make it useful.

Antonio pointed to a log alongside the path. It looked as if it would make a good bench for a tired doctor, but I had been in the jungle long enough to know that fallen trees were not for sitting. Peering closer, I saw the bark had sprouted hundreds of pastel-coloured mushrooms. Many of them were no doubt edible, but over, around, and between them, in constant motion, were thousands of 5 cm long, shiny black ants.

'Conga,' Antonio told me. 'Mucho dolor, mucha fiebre.'

I could believe it. Conga ants are ill-tempered insects with venomous stings very similar to wasps. Both carry their poison in their abdomens, which have pointed ends for stabbing. In fact, the two species are very closely related, with the same basic body design, except that somewhere along the way ants lost their wings, making them the foot soldiers while the wasps remained in the air force.

Antonio pointed out another ant that was smaller and less aggressive-looking than the conga. I didn't know the name but I recognized it. Indians crush it and then inhale the vapours to cure colds. I made a motion of grinding my thumb into my palm, then closing my fingers loosely around my nose. I wanted Antonio to know I remembered what he had taught me.

There was a neat black stripe across the path ahead of us, and as I got close I saw that it was alive with movement. Thousands of ants were crossing the trail in both directions. With the thick underbrush on both sides I couldn't tell where they were coming from or where they were going, but I knew what they were: army ants, the terror of the tropics. Army ants congregate in colonies of millions and are the only ants without permanent homes – they are too destructive to linger long in any one place. Every month or so, the queen lays a quarter-million eggs. With that many mouths to feed, the ants go on a rampage, dividing up into highly organized columns, each containing hundreds of thousands of soldiers searching for food. Army ants are carnivores. They'll kill anything in their path, from insects to crocodiles, then carry it back to their queen in bite-sized pieces.

Like all ants, in the tropics and elsewhere, army ants flourish only because they've formed a highly organized society. One ant would be powerless to affect its environment and unable to survive outside its protective 'civilization'. Humans have flourished in the same way. Pitifully weak and slow, we have increased to enormous numbers thanks to our ability to form societies. Without that 'artificial' protection, how many of us would survive?

But the protection of numbers is not truly artificial; it's a natural extension of the will to survive. One can imagine that single-cell organisms, capable of living on their own, started to stick together in pairs, and then in groups, to give themselves an advantage over their enemies and competitors. Each of the cells gradually specialized its design to allow the whole unit to function more effectively. The grouping of cells became what we see now as a 'body' – a collection of highly specialized cells working together and wholly dependent on each other's proper functioning for survival.

Human society is like an even larger organism – the next step up. The individual parts may not physically adhere, but the society operates as one animal, competing with other societies for survival through war, economics and culture. As is true with organ systems within a body, society functions best when roles are well defined and executed efficiently.

This particular society was on the move, but apparently not marauding; so we bent over to examine it. Both flanks were guarded by lines of larger ants with large white heads and pincerlike jaws that looked like ice tongs. On their tails were venom-filled stingers. These were soldier ants, the fierce warriors who lead the attacks. I had learned about the ants from two of our expedition entomologists who were studying them at Zancudo Cocha, and I had already seen many soldier ants up close in jars around the camp. There they were biological specimens. Here in the wild, they were formidable enemies, unnerving to see in such great numbers.

Though I knew a lot about the ants' destructive power, I didn't appreciate their medical value. That was about to change. Antonio focused in on one of the soldier ants, following its movement for a few seconds, his pinched thumb and index finger hovering above it. Then he flicked his wrist and very neatly picked it up, holding it between its pincers and its stinger. He showed me the ant and then his finger, unbandaged, which he had cut earlier that morning on the sawgrass. He turned the squirming ant upside down and stuck its head against the wound. The ant instinctively snapped its jaws shut, pinching the skin edges together. Antonio then twisted the ant, snapping off its body and leaving the head embedded like a staple. Three ants were enough to close the wound.

We turned our attention back to the trail. Closer to the river, the route became more heavily travelled by animals using it as a highway to the water. Not that I saw any – an animal could be one tree away and I wouldn't have seen it.

Lying low and blending in are essentials for animals of prey. But one animal breaks the rule spectacularly, for its survival depends on being noticed. They make themselves so obvious that even I was picking them out, though none was longer than 5 cm. Their colours were dazzling, with striking contrasts – one a brilliant blue, another red with yellow stripes, another black with a bright red head, still another orange-swirled. They hopped about in broad daylight, as if to say, 'I dare you to eat me.' They were poison-dart frogs, the worst meal anyone can ever have. Biting into one releases a powerfully bitter taste, followed immediately by numbness in the face and burning pain throughout the body. Spitting the frog out might or might not spare the predator's life. For the reckless animal or human that tastes one, the meal proves an unforgettable experience, if not a terminal event.

The frog's poison is produced by sweat glands on its back and exists purely for defensive purposes – the frog has no way to inject it. Like ordinary sweat, it is released when the frog is scared, or bitten into. The toxic brew evolved as a germicide to cleanse the frog's delicate skin, which, always being moist, provides an ideal growth surface for bacteria and fungi. The skin contains neurotransmitters, chemicals that naturally occur in nerves – but in concentrations thousands of times stronger than normal. When it comes in contact with an animal, the poison floods its nervous system, shorting out circuits and causing paralysis of nerves and muscles. Some frogs produce a brew so effective it can kill a human who simply touches its back.

The poison frog won't benefit much if it has to be eaten to kill its predator. In this defensive strategy, an individual sacrifices itself to ensure survival of its species. The postmortem revenge of each frog quickly teaches a lesson to any would-be predator to stay away from the frog's relatives and descendants. Had the frog made the sacrifice voluntarily and with awareness of what it was doing, as sometimes occurs in higher species, it would be called altruism.

Altruism has such an immediately negative impact on the individual that it might seem natural selection should have eliminated it long ago. Yet altruism is common among organized societies from army ant colonies to human civilizations. The reason is that the drive to survive comes not from the individual animal but from its genes. Genes are interested in preserving themselves, but so long as there are lots of copies spread throughout a species, the loss of a few is insignificant. If an animal's sacrifice protects the other repositories of its genes, that would be fine from the genes' perspective; they come out ahead. The closer the sacrificing animal's genes are to the animal it is sacrificing itself for, the more effectively the system works; and in fact, the impulse for altruism is strongest for the immediate family, then progressively weaker for other relatives, others of the same ethnic group and others of the same species.

The system has worked well for the small poison-dart frog. Antonio, like other Amazon predators, eats only large frogs, and even then only after removing the skin. He said he saves the small beautifully coloured ones he catches to make poison-tipped darts for his blowgun. Though so far I hadn't seen one, I knew I was in blowgun territory. A blowgun is a hollow bamboo tube with a splinter of wood placed inside one end. A sharp blast of air propels the splinter out the other end like a rocket. The blowgun is made entirely from local ingredients, but I thought it might now be obsolete. I told Antonio I was surprised he still used one.

'I like hunting that way,' he replied. 'Though using a shotgun is easier.'

The trail seemed to come to an abrupt dead end. Ahead of us was a wall of small plants and bushes, as well as an overturned tree whose base was taller than I was. The tree had fallen recently, leaving a hole in the canopy, but already the sunlight that penetrated through that hole had stimulated enough new growth to obliterate the trail. We worked our way around it, stepping over some branches and limboing under some others, moving through a 'no-touch' forest. I couldn't place my hands anywhere for balance – leaves have itchy coatings, branches have thorns and barks have biting ants. I was constantly afraid to anger some insect, make a snake think it was under attack or inadvertently grab a poison-dart frog.

The temperature and humidity had been rising steadily; breathing in was like inhaling the exhaust from a Laundromat. I wasn't tired, but I was sweating. My shirt couldn't have been wetter had I swam here. Antonio must have thought I needed a drink and a snack. He eyed a thin bamboolike canawaska vine and tapped it with his machete. Deciding he liked the sound, he chopped out a section that included two knots. He tucked it under his arm as we went on a little farther. In a small clearing, we stopped for a break but I didn't sit down. I noticed Antonio remained standing as well. The canawaska section was a hollow tube sealed at both ends by the knots. Antonio cut into one end and a clear liquid ran out. He offered it to me first, but when I hesitated he took a drink himself. I soon followed. It was fresh-tasting water.

While I drank from the canteen he had made, he stepped off the trail and, with three or four strokes, hacked down a 18 metre palm tree roughly 1 metre in diameter. He chopped out a section of the trunk near the top and split it open. He took out the centre and then split that open. Inside were little white shoots. He peeled them and gave me some. They were hearts of palm.

Antonio was an opportunistic hunter. He spotted a tree with lemony-looking fruits just beginning to ripen. With a stick, he beat the branches until a good one fell, then he cut it up and shared it with me. Antonio really enjoyed it, and went back for a second one. The fruit had a lot of uses, he informed me – as a tea it brought down fevers, and the women used it as a contraceptive. It was a sweet-tasting dessert, but it had a lot of pits. I spit mine out; Antonio put his in his pouch.

He was completely at ease and relaxed here. The jungle was Antonio's home. Although he didn't have much speed or strength by animal standards, he had knowledge; and knowledge was enough to provide him with a comfortable life.

Lunch over, Antonio nimbly and quickly wove a basket out of leaves and vines and placed it in the clearing. As we left, it started to rain – a real downpour from the sound of it, but I wasn't getting wet. Only after the tightly woven canopy was fully saturated did some of the rain start to drip through. As we neared the end of the trail, animal signs were becoming more obvious, at least once Antonio pointed them out. There were capybara tracks, peccary droppings and even bark scratched by a jaguar. Antonio spotted an agouti, a rodent that resembles a big, tailless squirrel. As I turned to look, a raindrop washed some bug lotion into my eye. By the time I rubbed it out, the agouti was gone. I could, however, see some light through the trees. Though the leaves continued to drip, the rain had stopped and we were approaching a large clearing. The sun was out, and just ahead now there was a wide patch of blue sky. The green corridor ended. I stepped outside.

The village didn't have a name. It was a grouping of five huts – probably a collection of related families. The dirt around the huts was smooth and cleared of all vegetation except for occasional patches of bright red flowers. Two women were sitting on the ground and skimming their machetes over the dirt, removing any irregular clumps as if preparing a clay court for a tennis match. Behind the huts were rows of banana trees and cornstalks that blended into the jungle. Children were playing with or chasing a menagerie of pigs, dogs and chickens. There are many such settlements scattered across the Amazon. I wondered if there were any in which the Indians didn't wear T-shirts.

Each neat, square wooden hut was built on a platform raised up on stilts. Chonta-wood poles at each corner supported a roof of thatched palm. Part of the platform had no walls, making it a porch, while the rest had walls of split bamboo rising only halfway up so that it was unnecessary to have windows for ventilation or light or to see an approaching stranger.

One was approaching now. The dogs barked and the children stopped playing, staring at me as if I were a zoo specimen. Antonio and I crossed under a rope strung between two huts on which laundry was drying and stopped at the base of the next hut. A pig was lying in the shade against one of the stilts. It had a festering sore in its side and four baby chicks sitting on its flank. As Antonio announced our arrival, I watched the chicks peck at the pig and eat off its fleas.

To get into the hut, we climbed up a thick log with sawtooth steps cut into it that was inclined against the porch. In the semidarkness within, a girl was sitting on the floor. I recognized her as the one who had listened carefully to the instructions I had given Berullio about his son, her brother. She smiled and offered us seats on tree trunk stools. While she spoke with Antonio, a four-year-old girl was feeding noodles to a two-year-old, and I glimpsed a boy balancing himself on the roof in order to add thatch. We were in the front or 'guest' section of the hut, separated from the private living quarters by a bamboo wall. From behind that wall came an elderly woman holding a decorated clay bowl filled with a milky liquid. This was chicha, a traditional alcoholic welcoming drink made from manioc, a potatolike root vegetable that grows practically everywhere in this part of the world. I took a polite sip; it tasted like sour yoghurt. The woman then took a drink herself, but instead of swallowing it, she spit it back into the bowl. Chicha is prepared by chewing the manioc to mash it up while adding spit to speed the fermentation. The woman did this continuously as we sat. I declined her offer of a second drink, though I convinced myself that the alcoholic content would kill the bacteria in the bowl and that my health risk was less than it would have been had I kissed the old woman on the lips.

The boy on the roof had been quietly weaving thatch the whole time we were drinking chicha. Now that we had finished, Antonio called him down. To my surprise, I discovered that he was my patient, Hermanigildo. No bandage, no splint and apparently no problem. I asked the sister how long ago they had stopped following my instructions. She said just yesterday, but patients will always say that no matter how long it's actually been. She had given me precisely the same answer I get all the time from non-compliant patients in New York. The surgery had been two weeks earlier. I wouldn't have been surprised if the bandage and splint had come off the next day. When I asked the boy about the bandage and the splint and the antibiotics I had given him, he looked at me blankly.

They had ignored all the advice I had given them. They knew better than I that my instructions, which were what I would have given a patient back home, did not apply here. The Indians believe in seeking cures, from witch doctors or otherwise, but societal pressure prevents them from doting on a wound. The exigencies of the jungle simply do not allow it. That kind of luxury can only exist in societies with much thicker protective barriers. In a small group, in which survival is a full-time job for each member, those unable to pull their own weight cannot be carried for very long without endangering the whole group. Individuals adapting to their environment must also, therefore, adapt to the rules of the group that provides their protection. The harsher the environment, the more stern and unyielding its rules; the group must do what is best for the survival of the species, even at the cost of the individual. The principle of preserving genes, which motivates altruism, also works in reverse to motivate baser instincts. To maintain the gene pool at large, it is sometimes necessary to detach a small portion when it drains a disproportionate amount of resources. In practical terms, an Indian boy has to learn to take care of his injuries himself because the tribe can't afford to do it for him for very long.There wasn't enough light in the hut for me to examine Hermanigildo's wrist, so we moved out on to the porch, though the light there wasn't much better; an aluminium pot in the corner held a smouldering fire that was filling the porch with blue smoke. I put on my headlight. Judging by the amount of dirt ground into the palm, the bandage had been off for quite a while, but the wound had healed well. The boy held his hand in a normal posture, not like a wounded paw, and he lifted his wrist readily. Children heal quickly; yet I felt something else was at work here. The need to survive generates a positive force within the body – actual physical changes that intervene to alter the course of a disease or to speed up healing. It's well known that positive thinking promotes health. People heal faster, live longer. Whatever influences the mind to generate that force – a placebo, religion, will, word of honour, desperate circumstances – the effect is the same: the ability to recover from injury is improved. It is a crucial adaptation to any extreme environment.

The sutures were ready to come out. Unlike the ant heads that Antonio had put in his finger earlier that day, these would not fall out by themselves. I took out my forceps and scissors to begin work. Then I had an inspiration. I took out the achiote paste left over from this morning when Antonio had insisted I spread some around my ankles. Before removing the sutures, I carefully painted my face with the same red lines Antonio had designed for me. In this setting, at this time, it seemed to fit, though no one reacted outwardly.

The sutures were removed and the wrist was working; it was still a little weak, but there is no Indian word for 'rehabilitation therapy', and anyway, the jungle would provide all the motivation and exercise Hermanigildo needed. I had repaired his broken parts, but I had not made them heal; that he had done himself. I thought there was nothing left to do, but I was wrong. The boy was waiting for the rest of his treatment.

Amazon Indians live in a natural pharmacy, and each tribe has found remedies for the diseases they all face. From tribe to tribe, however, the same botanical specimen may be used to treat very different conditions, and conversely, different plants may be used to cure the same disease. The reason lies not in the plant's chemistry but in the aura it creates for the patient. Even if the plant has no physiological properties, the belief that it does can give it real power. When that belief is reinforced by ritual, the treatment will be even more effective.

Antonio had gone outside while I was finishing up. He returned a few minutes later with a branch from which sap was seeping from the cut end and he dabbed it on the wound. Then he crushed some of the leaves in his hand, sprinkled the powder on to the sticky sap and rubbed it in. The boy had been waiting for that step; he left contentedly. His sister told me that Hermanigildo had been receiving this treatment from none other than Antonio, who had been making twice-weekly visits.

'He wanted the sap, and he wanted me to put it on,' Antonio said. 'Your pills are good too, but sap heals faster.'

I nodded my approval, though I felt the effect was due not so much to the ingredients as to the act of applying them. Treatments may vary, but the need to inspire confidence – whether the technique begins with a white coat or red face paint – is universal. Antonio understood this as well as I. He said, 'I can cure my people because they believe in me.'

Our work on the porch had attracted a crowd of children who were watching us from below, one of them steadily bouncing a tennis ball off his head. Like the cartoon stickers distributed by the nurses in my office, the tennis ball had gone a long way toward breaking down the fear all children have of doctors. This group seemed to be waiting for me to play ball with them. Then something more interesting caught their attention. They heard yelping, and ran off.

Having suddenly lost my audience, I went to see what the commotion was about. Antonio followed me. The children had gathered at the edge of the clearing, around a dog lying in the brush. The yelping had stopped. Tremors were passing through its body and combining to form waves of convulsions. It was bleeding from its nose. Then it vomited bright blood and stopped moving. The children quickly lost interest. Only Antonio and I remained at the spot. I asked him what had happened. He shook his head and replied that some of the children were careless. They had gone out hunting with their blowguns that morning, having tipped their darts with a dart-frog poison and curare. For every animal, the Indians know the specific mixture of poisons that kills fastest and best. The quantity placed on each dart has to be carefully measured: enough to bring the animal down quickly but not so much that it spoils the taste or makes the meat dangerous to eat. The part of the animal around the dart is cut out before eating because it holds too much poison.

Out hunting monkeys, the boys were unable to hit any, so they settled for a toucan. The bird was shot with a dart made for a much larger animal, hence its whole body was poisoned. The boys knew the toucan was inedible and threw it into the brush. But not far enough away from the hungry dog, which had tried to eat it. Parts of the mauled bird were lying around the dead dog. It was a gruesome scene, but an amazing display of the power a natural poison could have. That poison had been extracted from a living plant and a living frog, in which it had caused no harm, yet a small dose had been enough to kill a bird instantly and, after diffusing throughout its body, remained strong enough to kill within minutes the dog that tasted it. Antonio saw that I was intrigued.

'I'll show you how it is made,' he said.

He led me to the edge of the village clearing, but before we started our tour around it, he removed from his pouch a few seeds from the fruit we had eaten for lunch, dug a shallow hole and casually dropped them in. I looked around us and realized that what had first appeared to be a random collection of plants and trees was actually a peripheral garden, supplying not just food but medicines, fabrics and building materials for the families within. They had modified their personal space in the jungle. In a small way, they were adapting the environment to them rather than the other way around.

The garden served as the village pharmacy, a renewable storehouse of medical supplies; and Antonio, though not a shaman, was a repository of knowledge about medical and magical cures for the body and the spirit. As we walked, he pointed out plants used to relieve pain, heal infections and get rid of parasites. The bark of one tree on which he rested his hand was useful as a treatment for rheumatism; its leaves offered a cure for diarrhoea. Antonio explained that red saps were good as disinfectants, while white ones treated upset stomachs. (It struck me how the colours matched those of iodine and milk of magnesia.) I learned that the root of the gumbolimbo tree relieved 'women's pain', which I took to mean menstrual cramps, while its bark treated 'men's and 'women's pain' – urinary infections. There was a leaf used to overcome electric eel shock and another for stingray barbs. Antonio uprooted the bulb of a small tree and declared that when fed to a dog it improved the animal's ability to hunt. He smiled when he pointed to one delicate plant with long slender leaves. 'If you make this into a tea and serve it to someone you love, that person will fall in love with you.'

We walked a little way into the forest so that Antonio could show me an ampiwaska vine. Wrapped around a tree was a green woody vine with red berries, pink flowers and leaves that looked like those of a philodendron. How odd it was to see unassuming houseplants thriving in the jungle, but in fact this is where many of them come from. They're suitable to bring indoors only because they've had thousands of years to get used to high heat and low light. The conditions here are similar to those in a house, so the transition doesn't require any sudden adaptation, something no species is good at.

The inner surface of the ampiwaska bark, Antonio explained, is scraped off and mixed with water, then filtered through banana leaves. The resulting amber liquid is boiled down to a paste. This is the prime ingredient for curare – a useful but dangerous drug I was quite familiar with as a muscle relaxant during surgery; given in too high a dose it causes death by paralysing the respiratory muscles.

Antonio showed me several other plants that he uses in his recipe for dart poison. Each had a specific purpose: one made the poison enter the bloodstream faster; another made it stick better to the tip of the dart; a third appeased the spirit of the dead animal. As far as I knew, none of those plants had been laboratory analysed, but I did recognize one ingredient from my surgical practise – a leafy plant rich in coumarin, the chemical used to make Coumadin, a medicine that prevents blood clots. Antonio adds it to the curare because once an animal is hit, it ensures that its bleeding doesn't stop.

Antonio varied his recipe depending on which animal he was after, employing the leaves, sap or bark of anywhere from three to fifteen plants. What he considered most essential, however, was an ingredient that doesn't even come from a plant: frog poison. The children collect the frogs as needed by grabbing them with big leaves to protect their hands. They are easy to catch because they're not in the habit of hiding and don't have much fear of predators. Antonio said he'd been using a frog that he still had in the hut. I reminded him that he had told me he usually hunted with a rifle. 'Yes,' he admitted, 'but lately I've been unable to get any ammunition.'

A rifle is an intruder in this ecosystem – made of materials that do not grow locally and based upon technology beyond the comprehension of the people who live here. Like any invading force, it is capable of radically disrupting the environment, but it needs a supply line for ammunition and replacement parts, and so far the jungle terrain has limited its spread. The blowgun, on the other hand, is a natural outgrowth of the jungle. Given man's clever mind but weak physical abilities, combined with his need to obtain food using the materials around him, its creation was inevitable. It represents a renewable resource: the expertise to make it gets passed down through succeeding generations of humans, and the materials from which it is made are replenished by succeeding generations of plants. The blowgun is a weapon in harmony with nature. But rifles are easier to use and more powerful, and in a clash of cultures the deadlier weapon always wins out. It is just a matter of time until the older tool for food-gathering is replaced by the new.

Leaving his biological warfare section, Antonio wanted to show me another department of his pharmacy, though the plants involved were scattered here and there among the others, and some were quite a distance from the clearing. He led me to a thin interlacing vine that rose up around a tree trunk in neat, regular coils. 'Ayahuasca,' he said. This vine is well known to Amazon Indians, field botanists and lab scientists alike because of the alkaloids it contains.

Alkaloids are chemicals that plants have developed as weapons of survival. Plants cannot run from predators, but they can practise chemical warfare. The alkaloids in ayahuasca are powerful nerve poisons with startling side effects. A preparation from its bark is capable of producing intense visions in multiple colours. Its alkaloids prevent the breakdown of neurotransmitters – those substances that temporarily float across the gap between two nerve cells to allow an incoming electrical pulse to 'spark' across and continue on its way. They are supposed to break down immediately after the spark and re-form only for the next pulse. That way, one nerve stimulation leads to one signal transmitted. Should the neurotransmitters remain in place, however, they will continue to stimulate downstream nerves and so create a flood of sensations. One major neurotransmitter, serotonin, a naturally occurring substance in human brains, gives a feeling of well-being. Too low a level causes depression; too much leads to euphoria. Ayahuasca doesn't make serotonin, but its alkaloids prevent serotonin from breaking down after it has been used – recycling it, keeping it in the loop and allowing it to transmit far more signals than are actually arriving at the nerve. Under the influence of ayahuasca, the effects of natural outside stimulation – a sunrise, a birdcall or a waterfall – are multiplied, distorted and intensified into hallucinations. When artificial stimulations are intentionally added – ritual music, dances and frightening masks – the effects can be overwhelming.

In tribes like Antonio's, when the boys reach puberty they are brought to a sacred waterfall for their rite of passage. They fast for three days, ingest ayahuasca and potions, listen to the constant din of the waterfall and search the heavens for God. Antonio remembered seeing a sky filled with jaguar eyes, darting snakes and virgins hiding behind clouds. They spoke to him and he replied, speaking words no one could understand. Only then was he deemed ready to return to the village to take up his role as an adult.

Antonio described the effects of ayahuasca as making him feel he was part of a larger organism, part of the earth; he no longer sensed any separateness from the world around him; his wants, his fears, his sense of time and of self all vanished. This didn't sound like a chaotic reaction to circuit overload. What he was describing was a harmonious phenomenon, virtually identical to what has been reported throughout recorded history in many, many cultures, whether brought about by potions and rituals or by prayer and meditation. However different the means of creating it, the experience is universal, implying that it involves effects on the most fundamental structures of the brain and therefore reflects its innermost processes.

The brain is hardwired with circuits of incredible complexity but with particular stations responsible for specific functions. Outside threats are monitored by a primitive, deep-seated portion of the brain called the amygdala. When the amygdala is stimulated, we feel fear. Orientation to time and space is a more complex activity carried out by the parietal lobes at the top of the brain. The left lobe maintains an awareness of where the borders of the body are. This part of the brain is not well developed in the first year of life, which is why an infant, looking simultaneously at his hand and at a toy, cannot distinguish which object is himself. The brain's right lobe assesses the space around the body. Together the two lobes form our navigational system. An injury to this part of the brain might result in an inability to calculate the manoeuvres necessary to sit in a chair. Awareness of self is centred at the front of the brain, in its most highly evolved portion, the frontal cortex, a processing centre and the seat of attention, alertness and concentration. Absent or rudimentary in almost all other animals, it takes years to develop, and until it does, humans have no concept of themselves as individuals.

For someone to lose his fear, his sense of time and place and his self-consciousness, the brain circuits to the amygdala, parietal lobes and frontal cortex must all be interrupted. The interrupter is the hippocampus – a primitive brain part that meters incoming signals and, like a circuit breaker, shuts down circuits temporarily when they get overloaded. Magic potions that recycle neurotransmitters to keep nerves firing; rituals that bombard the senses with pounding drums, gesticulating dancers and endless incantations; ceremonies that generate fear with scary masks, animal sacrifices or other acts of violence – all push the circuits into overload. Ritual participants deliberately exceed the capacity of the brain centres to handle the input from their environment. The hippocampus flips some switches off, maintaining circuits essential to life, such as breathing and heartbeat, but shutting down stimuli to the higher centres that are less important for immediate survival. With incoming signals blocked, the frontal cortex receives no stimulation and loses its awareness of the individual self. The parietal lobes, like any navigational system, cannot orient themselves without external time and location signals. The left side can't determine any body limit, so it perceives itself as endless and intimately continuous with the rest of the world; the right side loses all outside reference points, creating a sense of eternal time and infinite space.

Prayer and meditation can have the same effect when the individual wills the frontal cortex to focus intensely and exclusively on one thought – God, or a mantra. All outside input is effectively blocked out just as if a switch were flipped, so the acolyte in quiet concentration can bring himself to the same state as the ritualist in frenetic activity.

Visions often occur during deep meditation, a phenomenon thought to be due to the internal quiet, which lowers the threshold needed to excite a nerve. Random bursts of impulses, similar to epileptic seizures, fire into various brain parts, including the temporal lobes – the section of the brain beneath the temples that controls conceptual thinking and language and image associations. These bursts are more likely to occur when nerve cells are already irritated, as they are in epileptics, some of whom often experience auras or visions. These electrical bursts can also be triggered by fear, fatigue, low oxygen and low blood sugar – all of them common conditions in survival situations. This may be how people in desperate circumstances see their loved ones before their eyes or hear God speak to them and suddenly find the motivation to survive.

Such explanations for spiritual transcendence hardly mean that religious experience is a trick of our brains. We don't – and perhaps can't – know from where the spark comes to excite the nerves.

Antonio wasn't concerned with any of this, but his pharmaceutical knowledge was more sophisticated than even he realized. 'Visions from ayahuasca alone are often dull,' he said matter-of-factly. 'I like to add leaves from this one.' He pointed to a small green bush. It was a viridis, in the same family as the coffee plant, and known to contain tryptamine – a stimulant like caffeine. Tryptamines prolong the effects of serotonin, which ensures that viridis goes well with ayahuasca. Botanists have a great deal of difficulty distinguishing many of its subspecies, but it was no problem for Antonio. He pointed out two different varieties and said, 'This one brings out red in the visions and that one brings out yellow.'

Tryptamines work when taken in combination with other ingredients but are inactive when taken by themselves orally. The Indians, without understanding the means by which these potions work (at least by our standards – they have their own explanations), have gotten around this problem in their preparation and use of another tryptamine-containing vine, the Virola, a member of the nutmeg family. The bark must be stripped from the bottom of the tree early in the morning before the sunlight strikes it and, presumably, heats up the resin, denaturing the tryptamine it contains. The resin is boiled and then dried, resulting in concentrated tryptamine. Pulverized and then snorted through a hollow bird bone, it is absorbed like cocaine, directly through the nasal membranes, bypassing the digestive enzymes that would interfere with the spirit communion the Indians seek.

Sophisticated agronomy was at work in the village garden, for both practical and spiritual reasons. The Indians planted chonta palms to supply thatching material for their huts, but the leaves grow high up and the thorny trunks are unclimbable. An easy-to-climb tree was always planted close to a chonta palm so that there would be convenient access to the valuable leaves. Also for convenience, villagers rooted certain hard-to-grow medicinal plants in gouges cut into the bark of other trees – a simple but effective form of grafting.

Other ecological considerations were more subtle and complex. Trees will tell you when new seeds need to be planted,

Antonio told me, but only if you listen carefully. Certain types of bushes will grow well if they are planted together. That sounded like symbiosis to me: plants with complementary mineral requirements sharing the same soil. Antonio had a different reason. 'It is because their spirits are simpatico.' No scientist would be rash enough to think he or she understands all the sources of energy in the universe; in the jungle, Antonio's explanation seemed at least as good as mine.

Antonio was a teacher in the purest sense. It was a privilege to be in the forest with him. He was about sixty-five years old and had lived his whole life here. The knowledge of countless generations had passed to him from a tribe with no written language. Precious lore exists only in his memory and in the memories of those like him. When they are gone, the knowledge will go with them.

'Are any of your children going to be doctors?' he asked me, quite unexpectedly. The animation in his face was gone. Still holding the last plant we had talked about, he added, 'My children and my grandchildren are not interested in plants. They want chain saws and outboard motors for their canoes.'

Antonio led me back to the hut. We would have to leave soon if we were to return safely to camp before dark. It would be a long trip, and Antonio wanted me to eat something before we left. In my honour they were preparing a special meal – a freshly killed paca, a large rodent, which was just then being held over the fire to singe off its hairs. The fire rose from a large metal box resting on logs in the front section of the hut. Once the hairs were burned off, the creature was dropped whole into a pot of boiling water.

Antonio emerged from the back of the hut, carrying a small Tupperware container that had several holes punched in the top. He opened it slightly. Inside was a beautiful poison-dart frog, blue with black spots. Antonio showed me some darts as well. He explained that a dart is made from a bamboo shard, which breaks off like a big splinter. The shaft is smoothed; fuzz from kapok seed fibres is stuck on the back end to serve as a feather. He pointed to a notch just beneath the tip that creates a weak point. 'When a person tries to pull it out,' he said, 'the tip breaks off inside, giving the poison a chance to work longer.'

The frog poison is applied directly by holding down the frog with a bunch of leaves and poking it with the dart. A white, frothy sweat forms on the frog's back and the dart tips are rolled in it. Various curare blends get added later. The poison remains active for a year. One frog in captivity was good for about fifty arrows, and this frog was about done. Antonio would soon release it and, once back to its usual diet, it would quickly regain its toxicity.

I felt happy for the soon-to-be-liberated frog, which would be looking forward to its next meal. I couldn't say I was. The paca was boiling away. The old woman was mashing a bowl of manioc into a stew mixed with spinachy-looking leaves. One of the girls went out to take down what looked like a heavy white doily draped over the clothesline. This was manioc bread, which had been drying in the sun along with the laundry. On a shelf were some dusty cans of food, a bottle of rum and a bag of rice, but they weren't being offered. I resolved to eat the meal no matter what, so determined was I not to insult my generous hosts.

The paca, apparently cooked, was laid out on a plate. I willed myself to take at least one bite, but when Antonio sliced the whole animal into sections, he gave me the prized portion – the head. I knew that in the Amazon, refusing a meal was considered highly insulting but I couldn't eat it. I had set my mind on eating some abstract body part, not a recognizable head. Revulsion welled up in me. My conscious will could not overcome my emotional reaction. I had heard stories in which refusing a meal in the Amazon had proven fatal for a guest, and had I been in that kind of a situation, I know I would have been able to eat the rodent's entire head. Even cannibalism becomes plausible when the situation is desperate enough. But survival was not at stake here, hence instinct could triumph over higher mental processes.

The family was disappointed. Antonio, however, having had

considerable contact with outsiders, was not surprised by my antisocial behaviour. He quickly withdrew the serving and replaced it with some manioc stew and a piece of bread. I was more than happy to eat that, while the others enjoyed their paca – boiled, I was assured, to perfection.

After the meal one of the younger girls brought out an English Bible and asked Antonio if I would read aloud from it. She didn't understand English, so it didn't matter what I read, her grandfather said; she just wanted to hear what English sounded like. I chose a page at random, read it aloud, and she was satisfied. In fact the whole family seemed pleased. It felt like the right time to depart, so Antonio and I both got up. The elderly woman motioned us toward the smoky blue fire that had been burning all day on the outside porch. Inside the aluminium pot was a smouldering termite nest. Antonio picked up a piece, crumbled it and rubbed it on his face and arms. I did the same. I imagined I was performing some ritual asking for safe return to the camp, but Antonio said no, the termite smoke kept mosquitoes out of the hut. The powder would help keep them off us on the trail.

Going was much easier than coming. Antonio had reopened the trail just that morning so there was very little hacking to be done. As we walked it seemed like a good time to bring up something Antonio had said at the village.

'Antonio,' I said, 'when you explained how to make a blowgun dart, you said it will break off if a person tries to get it out.'

He heard the question in my voice. 'When I was young, there was warfare between tribes, and even shrunken heads. The arrival of the church stopped all that. Still, that's the way I learned to make darts. It works the same on animals as it does on humans.'

Now that the subject was open, there was something else I wanted to know. 'Did you ever shrink a head?'

He continued to answer without quite answering. 'The head of the fallen warrior was chopped off here,' he said, pointing to the interval just below my first neck vertebra. He explained that a vine was passed through the mouth and out the neck, then tied to make a strap so that the head could be carried back for preparation. The scalp was split down the middle, and the skin was peeled forward as the skull was removed. The facial muscles and the muscles around the temples and jaws were scraped out. The incision in the scalp was then closed with a running suture made from a tough vegetable fibre; the lips were sewn shut in the same way. The neck hole was sewn with a 'purse-string' type stitch, so that it could be easily opened to fill the head with hot sand and then shut tight. As the heat shrank the skin, the facial features were moulded to create a fearsome look. The sand, once it cooled, was emptied and replaced several times until the head was reduced to the proper size. The technique required the skills of a surgeon, pathologist and sculptor, but when done properly, it prevented the dead warrior's spirit from avenging his death.

So vivid was Antonio's description that I wondered how much more he wasn't telling me; but I didn't ask, and he didn't offer. We continued on in silence.

We stopped briefly at the clearing in which Antonio had placed his basket as a marker. The cut stump of the heart-of-palm tree already had fungus growing on it. The fruit tree, with Antonio's prodding, dropped more ripe fruits to the ground. The basket may have been hastily made, but it was strong enough to hold the half-dozen fruits that he collected and carried back to camp.

We finished the trail portion of the trip fairly quickly and would have been in camp a lot sooner had we not taken so long to cross the lake. Antonio stopped paddling midway along to listen to the water. Every so often there came a slight gurgling noise, though from where it was hard to discern. Whenever he heard it, Antonio would paddle a little bit more and then stop again. He was chasing a fish, and he stayed at it for nearly an hour before giving up. I couldn't imagine what fish could be worth so much effort.

Antonio's fruits were a big hit at camp. Because of the low water level, the expedition was unable to resupply itself, and this made anything fresh really welcome. Sebastián, our liaison with the Indians, told us how lucky it was that almost all our supplies had made it through the first time. He thought it very curious that the only things the Indians said had gotten broken en route were the three bottles of rum he had packed as a surprise for us. I didn't tell him I had seen one of them on the shelf in Antonio's village hut.

The next morning I was awakened by a commotion and went out to investigate. An enormous red fish was hanging upright from a rope strung between two trees, its tail nearly touching the ground. It had to be at least twice Antonio's weight and half a metre taller. Antonio was standing beside it, busily cutting out fillets. It was a pirarucu, an Amazon catfish, that makes a gurgling noise when it surfaces to breathe air. Antonio had gone out on the lake at first light, waited for the fish he had heard with me in the canoe the day before and harpooned it. The fish would add at least two or three days to our food supply and, I was sure, provide several meals for Antonio's family. I understood why he had hunted it so patiently yesterday. I also wondered what other surprises we might be swimming with in that lake.

The extra food gave the researchers enough time to complete their studies. Over the next few days, I treated mostly minor conditions and slipped into a comfortable routine. There were, however, sharp reminders of where we really were and of the challenges to our survival lurking just outside of, and sometimes within, our camp. One morning, I turned my boots upside down outside my tent and banged them together before I put them on, to be sure nothing had crawled inside during the night. It's a routine we all followed – except this time something fell out. It was a scorpion – every bit as mean as it looks. The scorpion has a brown crablike body with seven 'legs' on each side and an upward-curving tail with a stinger and venom gland at the tip. The front two 'legs' are actually claws with which it digs in to hold on while it flails its tail overhead to stab venom into its prey (like one of my toes, for example). The venom is a nerve poison, which contains digestive enzymes that liquefy soft tissues, making them easier to consume. The hard parts are left behind uneaten. The bite is immediately and intensely painful, and the poison can be as deadly as a cobra's. By the time I got over my surprise, the scorpion had escaped into the underbrush.

Other insects were also becoming too familiar with our camp. A member of the native crew, barefoot and in his underwear, came running to my tent one night, shouting that he had been bitten by a tarantula. It had crawled into his closed tent. We discovered that it had come from a nest of tarantulas living on the underside of the floorboards in our latrine. It was an ideal location for them: well protected, with easy access to all the flies attracted to the pit just below. Big as a fist, a tarantula is covered with black hair and has half-inch fangs – frightening and dangerous but normally not deadly, which is unusual in these parts. The Indian's leg had a painful swelling, but there was no double puncture wound, which fangs would have made. Instead, tiny hairs were scattered all over the swollen skin. The tarantula hadn't bitten him, but had felt threatened (the feeling had been mutual) and reacted by rubbing its legs across its abdomen, flicking off hairs like arrows. My patient had been shot with hundreds of tarantula hairs. They were easily removed with duct tape, but the pain and itching promised to last for weeks. The perpetrator had escaped, but its victim took solace in knowing that in the jungle tarantulas often meet horrible ends when they in turn are attacked by wasps. The wasp injects a paralysing venom and then lays an egg on the tarantula. Alive but helpless, the tarantula becomes a fresh food supply for the hatching wasp. In the Amazon, there is no mercy in the competition for survival.

So far our camp had protected us well enough from the hazards around us, but as a jungle outpost it was becoming more and more isolated. The water level was steadily dropping.

Soon it would be impossible to get the canoes out of the lake and on to the river. We might still hike out, but if the canoes couldn't be floated through the channel, we would be unable to take out much of our supplies, not to mention all our equipment and specimens. And once we did reach the river, we would have to find other canoes to get us out. Lowering water level or not, it still rained every day. Our clearing had become a sea of mud on which our tents were practically afloat. Kneeling on the tent floor would create an imprint so deep that when I rose up I could hear a sucking noise from below. But this was the Amazon and Antonio said the season had been unusually dry. We would have to leave in a couple of days.

While the biologists hurriedly finished their projects, I continued to run my clinic outside on a mat of heliconia leaves. Antonio often hung out with me when he had free time and I now routinely solicited his advice on patient treatment. One of our group came to me complaining of burning pain when he urinated. He had an ordinary urinary tract infection, but because we were leaving soon and spirits were high, I thought I'd suggest another diagnosis.

'Were you swimming in the lake today?' I asked.

'Yes, but I wore my underwear,' he said, a little defensively.

'Sometimes that's not enough,' I said solemnly. 'It looks like a candiru is lodged in there.'

He guessed that I was kidding, but I could tell a slight doubt lingered in his mind.

'If we can't pull it out,' I deadpanned, 'the only treatment is to cut the penis off.'

With that the joke was up, but Antonio had actually once mentioned that rather radical treatment to me, and I still wasn't sure whether he had really meant it. In any case, I had been in the Amazon long enough to know that the best treatment for a urinary infection was gumbolimbo bark, scraped and boiled into a paste. Antonio said he would prepare some for my still somewhat unnerved teammate. I said I would supplement that with a course of antibiotics.

One of my last Amazon patients was an Indian who had hooked a piranha and cut his finger on its razor-sharp teeth. I could have sewn the wound closed, but specimen jars filled with ants were readily available. With our entomologist's approval, and with his tweezers, I removed a few army ants. Under Antonio's watchful gaze, I placed four ants, one by one, head down over the cut. After they bit into it, I twisted their heads off. The wound closed neatly and Antonio smiled his approval.

Two days later we left the jungle, hiking out because our weight would have sunk the canoes too low into the water. As it was, they were already loaded down with all our specimens, equipment and supplies. With a lot of machete work to deepen the channel, the Indians managed to float the canoes out, dragging them part of the way.

We met up with the crew at the river, and I said goodbye to Antonio. I admired my teacher, yet I felt sadness for him. A teacher's influence endures if his students turn into teachers. In the United States, surgical technique is passed down by the principle of 'watch one, do one, teach one.' I had watched one; I had even done one. But I was a dead end. Antonio's wisdom was in imminent danger of extinction because he had precious few people whom he could entrust to carry it on. In fact, Antonio himself was an endangered species. His whole way of life was coming to an end. His people had prevailed in an incredibly hostile environment by learning ways to transform threats to survival into means for survival. Now, suddenly, oil pipelines, logging roads and airstrips are destroying his habitat, injecting an outside culture that spreads like an infection. Antonio himself is contaminated – by T-shirts, rifles and gringos who pay cash. Workers, settlers and scientists on expeditions are introducing strange artefacts and an exotic lifestyle. Antonio and his people can't cope with that kind of antigen, because it comes from a society, an organism, that until now has been entirely outside their ecosystem. Like a virus the body has never been exposed to, it encounters little resistance and spreads rapidly. Tribes that have learned to protect themselves from headhunters, crocodiles and poisonous snakes have never been attacked by modern civilization. They have no defence against something that targets not an individual but the entire culture. In harsh surroundings, survival of weak species, such as ants or humans, depends on their members maintaining links to their society. Weaken those bonds and you threaten the larger organism that keeps them alive.

A society collectively builds a storehouse of knowledge, tribal wisdom, that provides the spiritual and practical rules and tools to compete successfully in an extreme environment. Across generations, group behaviour is modified in response to changing conditions, but neither the individual nor his society has the capacity to adapt when the change is sudden, extreme or unnatural. The injection of an alien culture into the Amazon is all three of those, and it will soon make the Amazon tribes extinct.

But what of the individual Indians? If Antonio's grandchildren are dispersed into an alien culture, can they retain their 'Indianness' – the collection of mental and physical adaptations that allow them to survive in the jungle – so that they can be reintroduced into the jungle at some later date, like zoo specimens released into the wild? In a culture without a written language, knowledge is stored only as memory; it only takes a gap of one generation to wipe it out entirely and irrevocably. Children of subsequent generations will have lost the tribal wisdom that was their species' most effective adaptation to their environment.

Amazon natives do have at least some survival adaptations built into their bodies. They live there and drink the water, yet rarely get malaria and are at least partially resistant to parasites. These strengths, shaped by natural selection, are an essential part of them that they would carry back to the jungle. But the Indians itch when bitten by mosquitoes, feel pain when stung by conga ants, are not resistant to crocodile bites and die from snake venom. The Amazon is a tough taskmaster. Most of their survival ability is the result of mental and physical toughness forged into each person as he or she grows up in, and is hardened by, the demands of the surroundings.

The genes you bring into the jungle are less critical for survival than the adaptations you make once you get there. The Amazon provides some surprising examples of an organism's adaptive response to environmental conditions. Remove a poison-dart frog from the jungle, and its poison will progressively weaken. Put it back, and the poison becomes strong again. The frog's offspring, if born in captivity, will lack the ability to produce poison. If put back in the jungle, would they develop the poison their ancestors had? No one knows.

In this respect humans may be analogous to the poison-dart frogs. Someone raised in the jungle will retain his ability to survive there even after being away for some time. If his offspring were born outside the jungle, however, would they still have an innate ability to survive there? Or, as seems more likely, would they be as vulnerable as a brightly coloured frog without any poison?

The survival adaptations I saw in the boy on whom I operated flowed from three sources, each endowing him with both physical and mental strength. His society, with its tribal norms and mores, provided practical rules as well as spiritual support. His body, with its generations of genetic adaptations, had evolved to increase his immunity and perhaps also reinforce the brain circuits that facilitate pain endurance and encourage purposeful behaviour under stress. His environment, with its relentless supply of abrasive agents, had toughened his body with calluses and muscles, and toughened his will.

The survival instinct is too abstract a concept for people who act on it every day of their lives, as do the Indians of the Amazon. Living so close to the edge, they don't have the luxury, or the need, to step back and examine its origins. The idea of studying it would seem frivolous, the machines used to do it incomprehensible. Pondering the nature of survival is left to those strangers who fly in and fly out of the only world these survivalists know.

I left the Amazon from a jungle airstrip three canoe-days downriver from Antonio's village. As I boarded the plane I immediately sensed tension inside. A nun was holding a sheaf of newspapers like a bouquet of flowers, speaking with a strangely flat effect to the person next to her. As I watched, she unwrapped the sheaf to reveal three blood-stained spears. The sight chilled everyone on the plane. She told us the spears were part of a fusillade of eighteen spears used to kill a priest and two nuns who had ventured into the jungle just 27 km from where I had been. The murders had occurred on the same day Antonio told me about his tribe's warrior past.

There are still some isolated tribes resisting any contamination of their way of life, exercising their will to survive by attacking what they perceive to be alien human predators. Though 27 km is a long way through jungle terrain, even the most isolated tribes cannot hold out for long. The Indians living there are barely aware of 'Ecuador,' having no concept of what a country is, but they are surrounded by it on all sides. As terrified as cornered animals, they react violently to threats they cannot understand. They have nowhere else to go. Once their habitat is gone, they will be too. And the rest of us will not even know what we lost.

There is an overwhelming profusion of life in the Amazon, and we know virtually nothing about most of the species that live there. Plants, animals and humans are inseparably intertwined in a fabric of baffling complexity. They are totally interdependent, yet they battle each other mercilessly to stay alive. Each species is a winner; all the losers are extinct. Unravelling the secrets of their success would lead us down a labyrinthine path into knowledge of nature's most fundamental laws of survival. And if we are smart enough to listen to the people who have been competing successfully there for twenty thousand years, we might understand what it takes to survive in the most competitive arena on earth.

HIGH SEAS

THE MOVING WILDERNESS

ON THE SIXTH NIGHT of a solo Atlantic crossing, Steve Callahan had just prepared a pot of coffee and was lying on his bunk, wearing only a T-shirt, when his 6 metre sailboat collided with a whale. Thirty seconds later he was waist-deep in water, as a torrential flow poured into his cabin. In total darkness, fighting back panic, he tried to cut away the ropes that held his emergency supplies, but the cabin filled too rapidly and he was suddenly submerged. Holding his breath, he made a few desperate, futile slashes at the ropes before running out of oxygen. He forced his way up to the top of the cabin, pushed open the hatch, and got to the deck, which was awash with ocean. Callahan somehow managed to free the life raft and inflate it. Making sure it was still tethered to the boat, he shoved the raft off and dove into it.

Shivering in the raft, Callahan had his first chance to think. Every one of his actions so far, from jumping out of the bunk to diving into the raft, had been initiated by a survival instinct honed by years of sailing experience. Now he had to make his first conscious decision. He had escaped, but without his emergency bag. He knew he would not survive long on the supplies in the raft. The boat was still afloat, though bobbing below the surface whenever a wave passed over it. It could sink at any moment, taking with it those vital extra supplies. Cool judgement helped control the burst of energy produced by fear. He decided to take the chance. He climbed back aboard the boat, 'feeling,' he later said, 'a strange sensation of being in the sea and on deck at the same time.' He dove through the hatch into the flooded compartment and groped in the dark for his emergency bag. The hatch above him slammed shut.

Twenty-five-year-old Lucien Schlitz and nineteen-year-old Catherine Plessz were sailing the Mediterranean on an idyllic voyage bound for the tropics. Caught in a sudden storm, their boat was tipped over by a giant wave, flooding the lower compartments. The boat righted itself and, though low in the water, continued to float. Schlitz and Plessz were not experienced sailors. They decided to try to ride out the storm in their 2-metre life raft, even though their well-provisioned 6-metre steel cutter showed no further signs of sinking. After throwing in a few meagre provisions, they jumped into the raft, still tethered to the sailboat, and pushed off. Riding the waves for over two hours, they did nothing further to help themselves; they didn't even tie in their supplies. The tethering rope snapped, and soon they drifted out of reach of their sailboat. The storm still raged; the waves grew more and more menacing. Their raft capsized and they were tossed into the sea.

Only very light winds pushed the *Albatross* as it headed from Mexico to the Bahamas, having already travelled 16,093 km in eight months. The 28-metre square-rigged sailboat was a self-contained school ship, offering a year of combination high school and high adventure for sixteen teenage boys and their five teachers. Most were below deck one morning when they heard a sudden roar. The boat had been struck by a wind shear, a violent downdraught that strikes without warning and has been known to crash planes into runways. The slap of wind heeled the boat way over, throwing books, plates and equipment all over the floor. The students lined up to climb the ladders to the deck, thinking more about the mess they'd have to clean up than about any imminent danger. But the boat continued to roll over. The students who had made it topside slid off the deck, spilling into the water.

Someone suddenly thrust into a cold, turbulent sea has no time to prepare himself mentally or physically. His first objective is to avoid drowning. However, confusion and fear will be his first reactions. Fear automatically stimulates the production of energy. How that energy is used depends on the conscious process of sorting out the confusion, transforming the flood of sensory input into a coherent status report that can be presented to the brain's higher reasoning centres. With experience and discipline, the energy generated by fear can be channelled into purposeful action. Without it, energy will explode into panic and then dissipate into useless or even counterproductive activity.

It is ironic that the sea has become so terrifying and alien to humans. It was from the sea that all life arose. That was a long time ago, however. Most terrestrial animals have long since lost their ability to breathe water. Once submerged, the average human has less than two minutes to get his nose or mouth back into the air to survive. The body's natural buoyancy will keep only the top of the head above water (the 'dead man's float'). But when you're drowning, it's hard to keep your mouth shut: sudden and unexpected contact with cold ocean water stimulates a deep reflexive gasp that forces open the mouth while still below the surface. You need to move to get your face out of the water. Driven by fear, however, such efforts induce hyperventilation – quick, short breaths that stoke the body's metabolic fire in order to speed up energy production. The longer and harder you struggle, the more the urge to breathe intensifies, and, as with the reflexive gasp, this can open your mouth at the wrong time.

Even if you are able to exercise the discipline needed to hold your breath while struggling below the surface, you will soon enough run out of oxygen reserves. Higher brain functions will be the first affected. Oxygen flowing to your cerebral cortex will diminish, and since that is the origin of voluntary control, you will literally lose your willpower. This is why you cannot commit suicide by holding your breath, no matter how much you might want to. The drive to breathe is so primal that when no longer suppressed by fading signals from above, it resumes its automated functioning – inhaling regardless of the consequences.

Sooner or later a struggling swimmer will start taking mistimed breaths and take water into his mouth. Water that is not spit out has but two places to go – down the oesophagus (throat) to the stomach or down the trachea (windpipe) to the lungs. The stomach is used to receiving swallowed water. When it starts to hold too large a quantity, however, it becomes distended, pressing against the lungs just above it. The lungs are already having a tough time taking in air; the last thing they need is compression from below, preventing them from fully expanding.

The stomach does have a defence mechanism to counteract distension: vomiting. While drowning might not be the right time for the stomach to deal with its irritation, vomiting is another reflex over which we have no control, and it worsens matters by forcing more liquid and solid into the mouth. Seawater and stomach contents that are not spit back out or swallowed have only one other place to go – down the trachea and into the lungs. The trachea has muscles at the opening that contract into a watertight closure when they sense a solid or liquid. That protective function is called the gag reflex, and it's what prevents food from 'going down the wrong way.' That's why it's a good idea not to eat and talk at the same time. That's also why it's not a good idea to have seawater in your mouth and breathe at the same time. A drowning swimmer, however, has no choice. The trachea opens for air and gets water instead. Maybe also some vomit. All this spills down toward the air passages. Like the stomach, the lungs have a secondary defence. Reacting to the invasion, they propel their reserve air supply upward to try to blow the stuff back out: the reflex of coughing. Coughing may be lifesaving on land. In the sea, it forces open the mouth, allowing even more water to enter.

Lungs are not gills; they are collections of thousands of membranous air sacs – tiny balloons that get pumped full of air with each breath. As the air comes in contact with the lining of these sacs, oxygen is extracted and drawn through the membrane on to its outer surface. Each sac sits in a net of blood vessels that absorb the oxygen and load it into red blood cells that travel through the bloodstream like ore carts through a mining tunnel. The cells make a round-trip through the body, distributing their oxygen supply to the various organs along their route. At the same time, they collect carbon dioxide – the principal pollutant created by the body's machinery. When the red cells get back to the lungs, they refill with oxygen and at the same time off-load the carbon dioxide, which is expelled with the next exhalation.

The body's appetite for oxygen is voracious. Survival depends upon maintaining a high-speed, high-volume interchange between oxygen and carbon dioxide. Oxygen supply remains the weakest link in our support system, and yet surprisingly the alarm that warns of danger is not a low-oxygen detector. The overwhelming need to breathe that we feel after roughly a minute of holding our breath is generated by sensors that monitor carbon dioxide buildup, much as a canary in a mine shaft warns of poison gas accumulation; the canary stops singing when noxious gas levels rise. The body's sensors sing louder, on the other hand, sending electrical impulses into the hypothalamus, the part of the brain responsible for pacing respiratory muscles. The signal becomes more and more urgent as the carbon dioxide level rises.

Groping around in the dark of his submerged cabin, Steve Callahan was rapidly accumulating carbon dioxide and receiving urgent signals to breathe. With conscious effort, however, he was able to override the impulses that would have automatically restarted his breathing – and drowned him. He was able, temporarily at least, to dampen the signal by generating a countercurrent originating in his frontal cortex, the seat of his will. During that minute or so between the alarm going off and the start of brain cell death – known as the lucid interval – purposeful movement remains possible. Callahan managed to find and cut loose his emergency equipment bag, then, encountering the shut cabin hatch above him, was able to force it open. Finally, gasping for breath, he reached the surface. Later that night he watched from his raft as his sloop rolled on to its side and disappeared beneath the waves. He no longer had a boat, but he had his emergency bag. Productive use of his lucid interval would save his life.

Lucien Schlitz and Catherine Plessz were also trying to stay alive. Only a few hours earlier they had been dreaming of the tropics; now they were two bodies in an alien environment. Since the capsizing was not a complete surprise, and the waters of the Mediterranean are not especially cold, they had avoided the gag reflex that would instantly have drowned them. Once past that danger, humans underwater will automatically hold their breath, an instinctive reaction. How deep-seated it is I demonstrated to myself the first time I put on scuba gear. As soon as I submerged, I stopped getting any air, so I surfaced and switched tanks. My second rig proved no better. When I explained the problem to my instructor, he told me that the blockage was not in the hose; it was in my head. What was preventing me from breathing was instinct.

Their life jackets kept Schlitz and Plessz's heads above the surface, and due to instinct, they hadn't yet swallowed any water. Pounded and pummelled by heaving waves, they held on to their overturned yet still buoyant raft. With waves washing over them continuously, their air intake contained large quantities of water, but being conscious, they were able to cough back up any water that entered their lungs through mistimed breaths. Preprogrammed automated survival defences, and technology, were keeping them alive. So far.

Between waves, they somehow managed to right the raft and climb inside. Their supplies, which they had neglected to tie down, had been washed overboard. Under relentless attack from wind and waves, the raft flipped over again. Once more they were able to turn it right side up and get back inside. The wild ride continued all night, tossing them into the sea a third time. Again they struggled back on board. This time they held on until morning. Huddled in their empty raft, with no food or water, they looked out at the now calm sea, powerless to regain their well-provisioned and still floating sailboat, which had disappeared over the horizon.

Also still afloat was the *Albatross*. Though it had been pushed on to its side by a sudden wind shear, it was still within reach of the students who had slid off its deck. They were relieved to see the boat right itself. Then, horrified, they watched it sink rapidly. Its lower compartments had been flooded. Some of their classmates and teachers were still inside.

A lungful of air will provide at most about two minutes of consciousness. Frantic activity and fear will burn it up a lot quicker. Even two minutes is not enough time to work your way out of the maze of underwater passages in a 28 metre sailboat plummeting to the bottom of the ocean. As the air supply runs out, the metabolic fire dims, and like any fire burning without enough oxygen, by-products build up. Chemical reactions in every organ are disrupted. The brain, critically dependent on a steady supply of oxygen, is affected first. It selectively protects the lower, more primitive centres, the ones that maintain heartbeat and breathing, at the expense of the higher, less immediately critical centres such as the ones that control reasoning and will. Blood containing what little oxygen is left is shunted to the hypothalamus and away from the frontal cortex, in a desperate attempt to maintain basic life functions. The chemistry and electricity of the frontal cortex begin to go awry. Confusion and panic overtake reasoning, and willpower cannot be generated. The insistent signal to breathe goes unchecked. The mouth opens. Seawater rushes in, and what is not swallowed runs down the trachea into the lungs. The mechanics of breathing are violently disrupted. Air sacs fill like water balloons. Oxygen cannot be extracted, and carbon dioxide cannot escape.

Making matters even worse, the seawater sloshing around inside the lungs is 'washing' the lining. Air sacs have an elastic coating on the inside so that, like any balloon, they can contract when the air goes out. But once that coating is rinsed away by seawater, the formerly springy balloons turn into flaccid bags. Water flows in but not out. The lungs become heavy and sodden, like a wet sponge. Those air sacs that are still dry are unable to push themselves open against the soggy mass. Without gas exchange, we lapse into unconsciousness, and when that happens, we lose our last defence against the sea – the gag reflex, which closes the trachea when there is food or liquid in the mouth. The gag reflex works only in conscious individuals. That's the reason why patients having anaesthesia can't have anything to eat beforehand and why rock stars who overdose have been known to choke on their own vomit. An unconscious person taking in water can't keep his trachea closed. The floodgates open, water pours in and the lungs are finally overwhelmed.

Brain cells can endure only four minutes without oxygen before they start to die. This was not enough time for the six trapped teachers and students on the *Albatross* to get to the surface. It was enough time, however, for Steve Callahan and for Lucien Schlitz and Catherine Plessz, who accomplished the first goal of any shipwreck victim – not to drown. The second is to get out of the water. When the *Titanic* sank, the survivors were the ones who made it into the rowboats. No one in the water lived, even though many wore life jackets and the water was actually warmer than the air. They died because water conducts heat away from the body thirty times faster than air. The water sucked the life out of them. Being on the water is far preferable to being in it, though those lucky enough to make it into a raft now confront the combined assault of sun and sea and get the chance to die of thirst, hunger, exposure or isolation. Unless they are eaten by sharks first.

The ocean is not only the largest wilderness on the surface of the earth, it is the only one that moves. Everything that floats washes ashore sooner or later. Currents, tides and winds ensure that anyone lost at sea, unless he or she is eaten, will eventually reach land or be spotted by a ship. If a human being can withstand the stresses of the ocean long enough, he or she will be rescued. Practising extreme medicine on the high seas means sustaining a human body long enough so that when the life raft washes ashore, there is still life in it.

Before the whale, the storm and the wind shear, the solo sailor, the romantic couple and the high school students were comfortably settled on their boats. A few minutes later, they all found themselves adrift in life rafts – cold, wet and in shock. The *Albatross* happened to sink in the Bahamas along a shipping route; the following day the surviving students and teachers were picked up by a freighter carrying a circus to Florida. When Lucien Schlitz and Catherine Plessz abandoned their boat, they were in the middle of the Mediterranean. A dozen ships passed them by before they were finally rescued two weeks later. Steve Callahan's boat crashed into that whale off the Atlantic coast of Africa, however, leaving him adrift in the north equatorial current with the nearest landfall 2,897 km away in the Caribbean.

Oceans cover three-quarters of the earth. The Atlantic is half the size of the Pacific, but that nonetheless represents 89 million square km. Callahan was floating on it in a 2-metre-long raft. Military satellites can photograph the sea with enough resolution to pinpoint a life raft. Scanning just the Atlantic, however, would require one trillion photos. And since he did not have the time to send a distress radio signal, Callahan knew that no one would know where to look for him. It would be weeks before anyone even realized he was missing.

Though humans can survive only a few days without water, most people adrift on the ocean do not die of thirst. They don't survive that long. They succumb to ignorance, fear and despair – or to seasickness. Nautical and nausea share the same Greek root. Seasick sailors spend the first half of their voyage afraid they are going to die and the second half hoping they will. Seasickness, however, is no joke. For a castaway it can be fatal.

'Harvey staggered aft, doubled up in limp agony. His head swelled; sparks of fire danced before his eyes; his body seemed to lose weight. He was fainting from seasickness, and tilted over. A wave swung out and pulled him off and away. The great green closed over him.' Harvey is a character in Rudyard Kipling's *Captains Courageous*, but his demise is no fiction.

Our bodies determine our orientation through three lines of information. Eyes seek out level surfaces to ascertain if we are tilted. Position sensors in bones and joints feel gravitational pressure, indicating our posture. Fluid-filled tubes in our inner ears change flow direction in response to acceleration, turning and spinning. Taken together, they give us a good idea of our position and whether we're moving – so long as all the information coincides. Even an experienced sailor can get seasick when he suddenly finds himself in intimate contact with the sky, the sea and a raft, and all three are providing sensory input that doesn't match. His eyes see an unmoving background of sky and horizon but a foreground of undulating waves. Those waves, in turn, are out of step with the signals from the body's position sensors, which are experiencing the waves directly under the raft. Also, the raft is rolling, pitching and veering, churning the inner ear fluid in three directions at once. Taken together, it's enough to make you sick.

The nausea might be relieved by vomiting, but that creates more problems than it solves because it deprives the body of food, water and electrolytes – minerals like sodium, potassium and magnesium that are critical for nerves and other body tissues. Seasickness increases energy demand – vomiting takes strength – but decreases energy supply. The hapless victim becomes weak and dehydrated and, lacking the essential ingredients for nerve transmission, loses dexterity, judgement and the will to live.

The first few days on a life raft are critical for survival. A lingering case of seasickness can mean the difference between life and death. Drugs such as scopolamine or Dramamine – if your raft happens to stock them – will provide some relief. If not, acupressure might help. Pressing two fingers into the centre of your wrist stimulates the median nerve, the main nerve of the arm. This sends a steady impulse to the brain that interferes with the discordant signals it has been receiving and makes them easier to ignore. It is also a good idea to avoid tasks requiring a tight focus, such as stowing food, making the raft seaworthy and taking navigational bearings, but that would mean avoiding precisely what is necessary to stay alive. Swimming provides an antidote to seasickness, though in the open ocean sharks are everywhere, and they are attracted to limbs dangling in water, especially if the water has already been chummed with food particles previously vomited overboard. Sometimes the treatment is far worse than the disease.

There are many other cures for seasickness: deep breathing, cassava beans, ice on the neck. I myself once found that a wet teabag placed on the forehead was highly effective. What these treatments have in common is the placebo effect. There is no medical reason for any of these to work (at least not as far as we know), but if the sufferer believes it will work, it will lower anxiety. Reducing the fear that is the natural reaction of any castaway will allow the brain to control the conflicting signals. After a few days, the brain generally manages to sort out the information from the eyes, joints and ears. It learns to rely mostly on the fluid in the inner ear, which functions like a carpenter's level to keep the body in balance. The sailor regains his sea legs, provided that his body has not already been irreversibly debilitated – or washed overboard.

The ocean is home to 90 per cent of life on earth, but for the 10 per cent that can't drink seawater, it's a tormenting desert. Castaways like Steve Callahan have three to six days to find fresh water before they become too weak to drink and too delirious to remember why they wanted to. A body can last weeks without food but will dry up in nine to twelve days if it is not watered. Callahan was thirsty almost immediately. He had swallowed a lot of salt getting his supplies and equipment into his raft. He had enough food and water on board to last two weeks. However, he calculated that it would take about three months to make landfall in the Caribbean. He knew that fresh water would not be coming from the skies. His raft was being pushed across the Atlantic by a trade wind that blows from the Sahara. All the moisture had been baked out of the air, which has to cross half the ocean before it will pick up enough evaporation to give back rain. He couldn't count on a gift from heaven.

What he could count on is what his species has always counted on for survival – applied technology: the innumerable inventions that have allowed humans to make their way into environments to which they cannot adapt physically. On board with Callahan was a solar still, a simple but delicate piece of equipment that turns saltwater into fresh. There were also some fishhooks, line and a speargun. And there was the raft itself, a square of inflated rubber tubes partly covered by a domed canopy with just enough headroom for him to sit under. The thin floor sagged around his knees and rippled like a water bed whenever he moved, but the roof blocked the sun, and the sides kept out the sea. The still, the fishing gear, the raft and the skills to use them were a legacy that creative minds had passed on to him. Now that he was completely isolated from the rest of his species, his survival as an individual would depend on how he used that legacy.

Without the raft, Callahan would have been eaten by a shark or drowned within a few hours; without the still he would die of thirst in a few days; without the fishing gear he would die of hunger in a few weeks. He was protected from a vast, powerful, natural ecosystem by a tiny, fragile, artificial environment, and he would survive only as long as he could preserve it.

The raft required daily upkeep. Air in the chambers expanded during the day when it was heated by the sun. Automatic safety valves relieved the pressure on the walls. When the sun went down, the air cooled and contracted, and the sides of the raft sagged. Each chamber had to be topped up with a bellows pump every night, and more often when there were leaks. The source of a leak couldn't always be found; even if identified, it might not be reachable, and even if reachable, it might not be patchable. Leaks under the water line brought the sea in from below; and a sudden incautious movement that enfolded a raft wall would bring a flood in over the top. Frequent, if not constant, bailing was necessary to prevent the raft from filling like a bathtub.

Sitting on the floor in a minimum amount of water, Callahan spent hours and days trying to set up his solar still. It turned out to be a finicky device. The still consisted of a plastic balloon with a seawater-soaked black cloth inside. Heat from the sun evaporates the water, leaving the salt behind. The warm water vapour rises and hits the top of the balloon, which is cooler than the superheated black cloth. Cool air can't hold as much moisture as warm air, so the vapour condenses out on the inner surface of the balloon. It is the same phenomenon that occurs when body heat fogs your glasses. The vapour forms into droplets of fresh water that run down the inside of the balloon into a collecting chamber. The system does provide drinking water, though it works much better in theory than it does in a raft.

Callahan didn't have the jaws or hands to catch fish, but with a fishhook and a speargun a human can become a predator of the sea. After twelve days, however, Callahan had succeeded only in distributing some of his own food as snacks to the fish. His reserves were meagre enough without having to share. Since the shipwreck, he had eaten just 1.4 kg of food. Only four more days of rations remained.

He wasn't having any success with the speargun either. There is more to spearfishing than meets the eye. What actually does meet the eye is a beam of light conveying the image of a fish on it. That beam is deflected as it leaves the water, projecting the image farther away than the fish really is. Like objects in a rearview mirror, fish in the water are closer than they appear. This is a problem that has long since been solved by cormorants and Eskimos. As a cormorant tracks a fish from the air and then dives in after it, its eyes undergo an optical correction that matches the refractive change in the water and keeps the fish in sharp focus. Not having the advantage of cormorant eyes, Eskimos solve the problem in a typically human way. They apply intelligence to experience and learn to throw their spears at empty water, aiming closer to the boat than where the fish appear to be. The Eskimos developed a way to compensate for their inadequate physical adaptation. Or, perhaps because they have the ability to develop skills, they had no need to develop specialized anatomy.

Having neither a cormorant's eyes nor an Eskimo's experience, Callahan was at a distinct disadvantage for open-water hunting. After many misses, he learned he should concentrate on the circle of water directly below his spear. Image deflection increases with distance. He realized that aiming straight down would minimize the distance so that the image and the fish would almost coincide. This was the kind of problem-solving under stress necessary for survival. Still, the fish wouldn't cooperate. Few swam under the boat. Leaning motionless over the side, his speargun cocked, Callahan spent hours stock-still, as if posing for a statue of a warrior.

Yet the sea is not an implacable enemy. A raft casts a shadow. To the animal and plant life below, the raft represents an unmoving island, for it is carried along by the current at the same speed they are. The bottom of the raft quickly becomes colonized by barnacles, then by molluscs and weeds. Within a few days, it becomes a nursery for worms, tiny crabs, shrimp and other crustaceans. Small fish gather beneath the raft and poke at the hors d'oeuvres on its undersurface. Larger fish, attracted to the collection of small fish, begin to follow along. Within a week or two, a marine ecosystem will develop – a food chain linked to the boat, literally within reach. Barnacles attach themselves so regularly to the bottom of any boat that sailors used to think they originated through spontaneous generation from the hull. Scraping them off is easy, and once they're shelled, they're edible. Combining them with the other denizens of the undersurface makes a sea mix that smells awful but doesn't taste as bad as it looks. Nevertheless, for long-term survival at sea, a human has to catch fish.

Activity began picking up under Callahan's raft. Large fish were starting to cross his narrowly circumscribed hunting zone. By the thirteenth day he had missed so many times that he could no longer concentrate. Absentmindedly, he took a pot-shot as another swam by, and to his complete surprise, he hit it. The 1-metre fish thrashed wildly as he brought it aboard. At all costs he had to prevent its snapping jaws and the impaled spear tip from puncturing the raft's air chambers. He stabbed the fish in the eye, hoping to paralyse it, but that made the creature's motions even more violent. Desperate to protect his raft, he wrestled with the fish, finally cracking its spine.

Callahan had won the grim death match. Now he prepared to eat his opponent. His survival instinct, intensified by hunger, had turned him into a vicious killer, but now, as those primitive impulses subsided temporarily, he was overtaken by a higher reaction: a sense of guilt for having brutally destroyed such a beautiful creature – a fellow mammal, a dolphin.

The capacities to feel guilt and appreciate beauty would seem to be odd evolutionary developments. Like other higher feelings, such as sympathy, mercy, duty, loyalty, honour and self-sacrifice, they seem hindrances to survival. But natural selection has fostered the development of noble feelings. They appear in higher animals along with more obviously valuable brain functions such as reasoning and memory, and it is unlikely they developed accidentally. It may be that they aid survival by endowing the will with something to stimulate it, intensify it and make it work both for the individual and the species.

It worked for Dougal and Lynn Robertson who, with their

twin boys, teenage son and teenage son's friend, were adrift in the Pacific after their 13-metre schooner was attacked and destroyed by killer whales 322 km west of the Galapagos Islands. What kept them going was what Lynn called their need to 'get the boys to land.' Responsibility provided a motivation stronger even than that of self-preservation. When latent primitive impulses are drawn up alongside the noble feelings, the combination creates formidable means for survival.

After seven days in a lifeboat, the Robertsons thought their ordeal was over when they spotted a freighter heading toward them. They fired signal flares, waved and yelled as loud as they could, then fell silent as the huge ship passed by. Up to then they had been anticipating rescue – a somewhat dependent mental state. The bitter disappointment ignited in them a fierce will. 'To hell with them,' Dougal said. 'We will survive without them.'

Knowing that their fates were in their own hands transformed them. Dougal felt 'strength flooding through me . . . and the aggression of the predator filling my mind. We had brains and some tools. We would live from the sea. From that instant on, I became a savage.'

First they needed water. Their supply was running low and they had no solar still. The skies overhead were cloudless, but they were drifting westward into the Doldrums, an area of the Pacific where there is more frequent rainfall. It wouldn't do for them simply to hold their mouths open when it rained. They would need to use every bucket available, including the one into which they urinated. The best rain catcher, they knew, would be the canopy over the raft. If properly shaped, it could act as a funnel, collecting water and channelling it into a container.

And they would need food. Attempts at fishing had so far been fruitless. Fish were collecting under the boat, but only small ones. In any case, they would not be able to catch large ones, since they had neither spear nor gaff. With his knife, Dougal was able to stab some of the fish, but the amount of food they provided was barely equal to the calories expended in catching them, not to mention the additional energy lost in anxiety over the possibility of losing the knife overboard.

The same day that the freighter changed them from passive victims to aggressive predators, another ocean traveller provided them with their key to survival: a turtle got entangled in the line of the sea anchor they were using to stabilize the raft. They quickly tied its flippers and heaved it upside down into the boat. Dougal pulled on its beak, its claws thrashing helplessly, and slashed its throat, spurting blood everywhere. After its shell was hacked off and insides sectioned, the 36 kg turtle was reduced to about 6 kg of turtle meat. Washing the blood off his hands, Dougal knew that before the freighter he would have recoiled at the thought of what he had just done. Now he would do whatever it took to keep his group alive. He was determined to create a niche for them in the ecology of the ocean. Yet when man enters an ecosystem, he doesn't necessarily enter at the top. Dougal and his family were somewhere in the middle. He cleaned his hands in the bottom of his boat rather than over the side; he was afraid the blood from his hands would attract a creature higher up the food chain.

The shark is the most efficient killing machine ever created. It has more ways of seeking out and destroying a target than a cruise missile. Sharks have been around for 400 million years – they're twice as old as the dinosaurs. Early models were over 15 metres long. Like many machines that have undergone years of development, their size has been reduced while their efficiency has improved. Modern versions are up to 6 metres long and can weigh over 1,814 kg – longer and heavier than most cars.

Sharks are not intelligent; they don't have to be. They have a remarkable collection of sensory guidance systems that allows them to home in on a target. They can be alerted by sound more than 2 km away and can smell a drop of blood or urine that's been diluted 100 million times. Their nostrils function independently, so by turning their heads they can constantly correct their path toward the smell. Once they get within a few metres of their target, the lateral line system comes into play. This consists of a series of fluid-filled holes along either side of the body that detect vibrations, particularly discordant ones, such as motion from a struggling fish or an uncoordinated swimmer. This works in tandem with a low-light vision system in which a layer of silver crystals at the back of the eye reflects incoming light so that it crosses the eye twice, thereby increasing contrast and perception of motion. At a few metres from the target, electrical detectors are activated. These are jelly-filled sacs in the head called ampullae of Lorenzini, which are extremely sensitive to the current generated by muscle contractions: they can detect the voltage in a single nervous twitch and even, perhaps, in an impulse of fear.

Sharks can stimulate quite a few fear impulses, as they did in my dive partner and me when we were diving off the Galapagos Islands while conducting a fish survey. I suddenly found it too dark to write on my counting slate. Looking up, I saw a shoal of hammerheads passing overhead, obscuring the light. One curious shark swooped down and bumped my partner. This is something they often do before feeding, much as a choosy shopper might test the consistency of a melon. We placed ourselves back to back, ready to defend ourselves. The shark, however, left to rejoin its companions. Either it wasn't hungry or it had determined that my partner wasn't ripe.

A shark's behaviour may be unpredictable, but its potential for destruction is unequalled in the animal world. Once it decides to lock on to a target, it arches its back and its normally graceful swimming style becomes erratic. The largest sharks accelerate toward the surface at up to 64 km per hour, striking with such force that they often drive their victim and themselves momentarily out of the water. A shark's bite is an attack by over two hundred razor-sharp triangular teeth arranged like rows of ripsaws inside jaws that close with the force of a bulldozer. The upper jaw has teeth 6 cm long for cutting. Teeth in the lower jaw slant backward for gripping. Sharks cannot chew, so they thrash their heads from side to side to tear off chunks of meat. The first strike may be just for 'sampling', though that may not be much consolation to the human victim, since the sample may be an entire limb or half a chest cavity. The shark may then decide the attack was a mistake, lose interest and swim off, leaving the victim to die from massive tissue loss and haemorrhage.

Sharks turn castaways into prisoners. They patrol all the oceans. The threat of an attack prevents a sailor from leaving his raft for a swim, confining him to a space smaller than a jail cell and denying him the exercise needed to counteract the physical effects of prolonged immobility. Skin will break down if it is in constant contact with any surface, whether a bedsheet or a rubber raft. The process speeds up when the skin is coated with a caustic liquid such as urine or seawater. In the case of people too sick to eat or unable to get food, their fat cushion disappears, and as a bone becomes more prominent, it can wear through the weakened skin. Debilitated patients who are not fed, washed and turned frequently will develop pressure sores, sometimes with bones poking out. Castaways who cannot exercise get the same sores – made all the more painful by the ever-present salt continually washing into the raw wounds.

And there are the mental effects: not being able to leave the raft can make the sailor stir-crazy, and the sight of sharks following the boat is a constant reminder of what will happen if the raft sinks.

A human's place in the food chain may be below a shark's, but it is above that of almost every other sea creature. Since all oceangoing fish are edible, humans can eat whatever they can catch. Spears are the best weapons. Resourceful castaways have made spears from knives tied to sticks and then attracted fish at night with a flashlight or by reflecting moonlight with a mirror. Fishhooks have been made from safety pins, compass needles, and fish bones. Biscuits or barnacles can serve as start-up bait; better bait can be made from the first fish caught. Anything from rope to shirtsleeves to shoelaces can be used for fishing line. Even with no bait, no hook and no line, a fish can be captured. One desperate sailor dangled his finger in the water and grabbed small fish that bit into it.

Marilyn Bailey, adrift in the Pacific with her husband, Maurice, for 117 days, thought of using her fingers to catch fish but then hit upon a less painful method. She cut a big hole in the side of a kerosene can, put bait inside and lowered it into the water by its handle. When a fish swam into the can, she simply scooped it up and deposited it in the bottom of the boat. Another way to scoop food from the sea is with a sock. Oceans are filled with tiny organisms – plankton and krill – that lie suspended in the water. They are the world's favourite seafood, preferred by whales everywhere. A nylon or silk sock trailing in the water will act as a filter to collect a serving of foul-smelling, repulsive-looking, but highly nutritious soup.

Thor Heyerdahl, the famous Norwegian anthropologist–explorer, used that basic technique and many others like it when he set out on a reed raft, the *Kon-Tìkì*, to prove that ancient peoples were capable of crossing vast oceans. One other food-gathering technique was to delegate the job to a small fish, called a remora, which attaches itself to larger fish by means of a sucker on the top of its head. When he caught one, instead of eating it, Heyerdahl tied a line around its tail and dropped it back into the water. The remora would soon attach itself to another fish with a suction strong enough to bring both on board when the line was pulled in. This ancient method is still used by Polynesians – an example of the human heritage of animal exploitation. The tradition was given immediacy by Marilyn and Maurice Bailey, who adopted it to stay alive.

Without a still or rain collector, or a spear, or a fishhook, or a fishing line, or even without socks, it is still possible to live off the ocean. In fact, the record for the longest survival at sea, and perhaps for the greatest ingenuity, belongs to Poon Lim, a man who had none of those tools but did have a fierce will to survive. Lim was a Chinese seaman serving in the British Merchant Navy during World War II when his ship was torpedoed and sunk by a U-boat in the South Atlantic. Floating alone on an open wooden raft, he managed to stay alive for 130 days. To provide himself with water, he ripped open his life jacket and allowed the canvas inside to become saturated each time it rained. To catch food, he made a hook out of a spring inside his flashlight and then made others by bending nails that he pulled out of the raft with his teeth. He unravelled some heavy hemp rope to make a fishing line and baited his hooks with biscuits. When he caught a fish, he cut it up with the edge of the biscuit tin.

Poon Lim was finally picked up off the coast of Brazil by a fishing boat. He had lived off the sea for over four months yet lost only 14 kg. He had to wait three more days before being taken to shore because he appeared so fit to his rescuers that they felt no need to interrupt their fishing trip.

Theoretically, no one should die of hunger on the high seas. There is a constant parade of fresh food. However, it is not easy to catch enough of it, and what you do catch will not make a balanced meal. A human diet should contain about 60 per cent carbohydrates, our preferred type of fuel and the one our bodies can't manufacture. We get them by eating plants or at least by eating animals that eat plants. But the wisps of algae that stick to the bottom of a boat are inedible, and ocean-travelling fish are not grazers. To maintain optimal physical performance, a human needs to eat about half a kilogram of carbohydrates a day, yet the amount floating or swimming in the ocean is virtually zero.

Carbohydrates are pure fuel. They are made only of carbon, hydrogen and oxygen, three highly flammable elements that provide a lot of energy when burned. When formed into simple combinations, they are called sugars. Sugars burn like paper. They ignite quickly, burn rapidly, but do not last long – a quick burst. When the same three elements are combined in a more complex way, they are called starches. Starches burn like wood. They require more heat to get going, but once they catch, they burn a lot longer and provide a lot more energy – a sustained release. Carbohydrates have no function in the body other than to provide heat and power. The liver and the muscles store them as readily accessible fuel depots. Once these storage areas are filled to capacity, the overflow of carbohydrates is converted to fat by removing some oxygen and rearranging the same three elements into a denser pattern. Fat is a concentrated fuel that can store energy more efficiently for the longer term. Humans make fat from excess carbohydrates, but they also obtain it directly by eating other animals that have already done that storage job for themselves. Unlike carbohydrates, fat is essential to maintain vital body functions like blood clotting, hormone production, digestion and nerve insulation. The vast majority of it, however, is dumped into otherwise empty fat cells, which are present almost everywhere in the body, especially around the abdomen, hips and buttocks, where their storage is sometimes too obvious. For a given weight, fat provides twice as much energy as carbohydrates. Looking at it the other way, losing half a kilogram of fat requires twice as much work as burning off half a kilogram of carbohydrates. This is why eating fat makes you fat.

Not more than 25 per cent of dietary intake should be from fats. This is no problem for shipwreck survivors. They subsist on fish, which are less than 1 per cent fat and have no carbohydrates at all. Fish are food because they are made almost entirely of protein. For a substance to be considered food by the human body, it must contain carbon, the fundamental element of life. For it to be edible, it must also contain hydrogen and oxygen in a form that body enzymes can break down and recycle. Besides carbohydrates and fats, protein is the only other substance that fits those requirements. But protein also contains nitrogen, making it far more complex than the other two compounds. Adding that fourth element allows for the formation of a variety of building blocks called amino acids, which, depending upon how they link together, endow proteins with an incredible range of properties. Proteins are the basic units of genes, the main components of muscles and bones and critical parts of hormones, enzymes, antibodies and tissue-repairing cells. They have some function in every living cell. But as complex as proteins are, they contain a lot of carbon, hydrogen and oxygen and are still food.

Some protein intake is essential because the nitrogen must be captured and reconfigured to build the human brand of protein. Once the nitrogen is removed, the remainder can be used as fuel. However, too large an intake of protein will tax the body's ability to process and dispose of it. A normal diet contains about 15 per cent protein, but a total seafood diet is close to 100 per cent. Because a protein is so complex, it requires five times more energy to break down than a carbohydrate, and the unused nitrogen needs to be highly diluted so that it can pass through the kidneys into the urine. Feeding exclusively on protein forces the body to use up large quantities of energy and water, an especially extravagant process for a hungry and thirsty castaway.

So the supermarket of the seas contains no carbohydrates and few fats, but it does offer an unlimited supply of protein swimming in variously shaped packages around the boat. For the castaway who can catch them, it is not a balanced diet, but it will keep him alive.

After a few weeks, Steve Callahan's raft looked and smelled like a fish market. Fish were not easy to catch, so when he did get one he was usually pretty hungry. First he cut out and ate the heart and liver – they can't be preserved, taste the best and are the most nutritious, especially the liver. Livers are chemical processing factories, filled with vitamins and minerals not present in any other organ. Cutting open the stomach sometimes provided a bonus of partially digested fish, which are edible once the bitter coating of digestive juices has been washed off. Next Callahan cut the muscles into thin strips, eating some right away since they are easier to swallow while still wet. The rest he suspended on strings hanging down from the canopy. If he didn't dry them in the sun, they'd spoil within hours. All the inedible parts he collected inside the boat, saving some for bait, not daring to throw the rest overboard until there was a strong current to carry the raft away from the sharks that the food would attract.

The Robertsons and the Baileys fed primarily on turtles. One 9 kg turtle supplies about 4 kg of protein and 36 kg of fat, so turtles are a better source of nourishment than fish. Besides the fat, there is always a good chance that the female will be carrying eggs – a much tastier surprise inside than half-digested fish.

Then there were the gifts from above that benefited everyone. Seabirds are attracted to a raft as a place to rest, and all the more so if they smell fish and blood. Once they alight and fold their wings, they can be caught by throwing a cloth or a shirt over them, then killed by suffocating them or twisting their necks. Peeling the skin off is easier than plucking the feathers. Eating them is not a good idea, though, for those who even half believe that seabirds are the wandering spirits of drowned sailors. The birds eat plankton, much of which is bioluminescent. At night, leftover bird parts have been known to glow like ghosts.

Another gift that arrives by air are flying fish, small fish chased at night, often by dolphins. When they sense they are about to lose the race, they jump out of the water and glide through the air, sometimes up to 27 metres, hoping to splash down in a safer spot. It is not unusual, though, that their desperate flight lands them in the bottom of a boat. Putting up a sheet, or some sort of barrier, in the middle of the raft each night greatly increases the odds of waking up to breakfast.

The greatest gift from the sky, however, is rain. A castaway might have all the food he needs, but if he can't drink, he can't eat. Most chemical reactions and body functions need to take place in and around water, a throwback to the days when life was lived in the primordial seas. The first chemical reactions of what became biology occurred in water, and though life systems have grown infinitely more complex in the aeons since, they have never gotten over their early origins. Humans may have left the sea, but they took their water with them. Two-thirds of a human body consists of water, distributed throughout every cell. The water is active within each cell and also flows from one cell to another cell, tissue or organ to facilitate body functions like digestion or circulation. In the process, some of the water is converted to other compounds, or lost outright, as in breathing, sweating and urinating. With no resupply, a human cannot function beyond five or six days or survive beyond ten or twelve. Rain remains the ultimate source of fresh water for all air-breathing animals. On land, it collects in lakes and rivers. A prisoner on the open sea has to collect it any way he can. A litre a day would be enough to sustain life almost indefinitely, but even that amount may be difficult or even impossible to obtain.

The Robertsons had six thirsts to quench, which they did primarily by catching rainwater in their canopy, used like a tarp to funnel the runoff into containers. Since the canopy was constantly exposed to salt air and to deterioration from the sun, the first 'wash' always contained salt and bits of yellow rubber. This spilloff was placed in a separate container so as not to contaminate the water that flowed down after it.

Not wanting to waste a drop, Lynn Robertson, a nurse, came up with the idea of using the undrinkable yellow water as a rehydration enema. Rectums have membranes that extract water – this is how food that has passed through the digestive system is dried into faeces. Transoceanic seabirds have a similar membrane in their throats to extract water. When the birds scoop up seawater, the membrane prevents the salt from passing through, so the birds get a drink of fresh water. Reasoning that the contaminated water would be filtered by the rectal membrane, leaving behind the salt and rubber, Lynn connected some rubber tubing to the container and to the bellows pump used to keep the raft inflated, then administered enemas to each of her family members. The teenage boy who was not part of their family refused treatment. Dougal considered ordering him to have it, but decided, on balance, that it was better for him to have less water than to force him to submit to something he found so humiliating that it might crack his spirit. Spirit is critical for survival.

Even without rainfall, humans can traverse the driest deserts if they can find the water hidden below the surface. Oceans also have oases of fresh water – small, mobile collections of drinkable fluid found within the body cavities of fish, turtles and birds. Sea life can provide at least some water if you know where to look. The blood of freshly killed turtles and birds can be drunk like gravy in the minute or so before it coagulates. An average turtle will yield about 4 cups, if collected under the neck when the artery is severed. Turtle meat and fish fillets also have moisture in them, and it can be extracted before drying by wrapping the pieces in a shirt and then wringing them out. Spinal cords are encased in protective tubes filled with liquid to absorb shock. Snap open a fish's spine near the tail and there will be a column of watery, sugary cerebrospinal fluid waiting to be sucked out. Eyes, another source, are filled with two transparent liquids, the aqueous and vitreous humours, which allow light to pass through from the lens, or cornea, in the front to the screen, or retina, in the back. Fish eyes can be eaten like grapes, chewing them one by one and spitting out the corneas.

Mining the sea for deposits of fresh water is difficult work, and under a hot sun the amount extracted might well be less than the amount of sweat expended. Castaways keep one eye on the skies, hoping for a bonus of rainfall, but water collection is limited to the capacity of whatever containers are on board. Not knowing when, or if, more rain is coming, they sip their reserves slowly, denying themselves the deep drink that would quench their unrelenting thirst. Then the next heavy rainfall might bring frustration that the containers can't hold any more – or fear that the storm might sink the boat. Castaways live from rainfall to rainfall, haunted by the fear of both too little and too much. And they have the additional torment of being afraid to drink any of the vast supply of water on which they are floating.

'Water, water everywhere, nor any drop to drink,' said the Ancient Mariner, and the US Army survival manual agrees. Yet many sailors have given in to the temptation to drink seawater. At first they just rinse their mouths with it, then they swallow some 'by accident'. It feels good in the throat. They lose control, taking sips, then gulps. Their thirst worsens, compelling them to drink even more. Soon they become delirious and irrational, and sometimes 'go for walks' overboard, providing another example of the sailor's adage that 'seawater sends you mad.' Second only to the effects of prolonged exposure, drinking from the ocean is the most common cause of death in a life raft.

But why should it be? We eat salt all the time, so why is drinking seawater lethal? The answer is that it changes the composition of the blood – a complex chemical mix that is about 1 per cent salt, compared with seawater, which is about 3 per cent salt. Blood is the only organ that is a liquid, and it flows everywhere, in contact with virtually every cell in the body. Chemical sensors continually monitor its ingredients and are very intolerant of increased salt, whether it comes from seawater, or pretzels, or anything else. Once dissolved in any liquid, salt is very hard to get out. Imagine pouring too much salt into a bowl of soup. The only way to get the taste back to normal is to add water to dilute the concentration. The body works the same way. The kidneys can filter some dissolved salt into the urine, but there is a limit to how much they can remove before they become corroded. As with oversalted soup, the most effective way to get the recipe of blood back to normal is to add water. In the body the process is called osmosis. Whenever sensors in the blood taste too much salt in their soup, they send a signal to the brain that makes you thirsty. This is why bars often put out free salted nuts. You find yourself drinking a second glass of beer, and when your salt content has been sufficiently diluted, you put the glass down.

If the salt concentration of seawater could be reduced to less than 1 per cent, any of it that was added to the blood would have the effect of diluting it and quenching thirst. Body sensors would get the blood's salt concentration back up by pushing water out into the cells, thereby rehydrating the body. Thor Heyerdahl and his crew survived an Atlantic crossing using this principle. They had unknowingly cut the reeds for their raft, the *Ra II*, in the wrong season, when the stalks were not water-resistant. As a result, the raft became waterlogged and sank to surface level. To prevent it from going under entirely, the crew jettisoned a lot of their supplies, including some of the heavy water tanks. This lightened the raft sufficiently for them to continue but left them short of fresh water. Given that the region of the ocean on which they were sailing was less than 3 per cent salt, they added 1 litre of seawater to every 2 litres of fresh water, increasing their water supply by 50 per cent – just enough for them to reach South America.

What happens, though, when there are no water tanks, the still is not working, the tarp is dry, even the cans and boots are empty and the rains don't come? Maintaining 1 per cent salt concentration in the blood is a high priority – vital chemical reactions depend on it – so if the body can't take in water from outside, it will draw upon its own supply, taking it from the cells the blood passes by. Within days, highly sensitive brain cells will demonstrate the effects of drying out as chemical and electrical systems misfire and short-circuit. A person with no water intake at all will lose consciousness in three to four days and die in seven to ten days. A person who drinks seawater will also die within seven to ten days but will remain conscious almost to the end. This is because the body naturally loses a little salt every day. If you limit the amount of seawater drunk to a pint a day, salt buildup will be slow, and body cells will absorb water. After about a week, however, the salt accumulation will overwhelm the kidneys and they will shut down. For a castaway, those extra three or four days of consciousness could be the time needed for fish to collect under the raft or for a passing ship to be signalled.

Whether the ship is there or not, dehydration can lead to

vivid hallucinations, my friend and fellow sailor Norman Baker recalled one day as we were served drinks before lunch at The Explorers Club. 'The last time I saw a cocktail waitress in a white dress like that was on a raft in the middle of the Atlantic.' Norman was the second-in-command on that *Ra II* expedition that jettisoned most of its water supply to stay afloat. At first Norman's waitress was just a recurring dream, but one hot, windless afternoon he was on deck dully splicing rope when she suddenly appeared alongside him. She offered him a glass of cold water. He was tempted to take it. He had been at sea forty-three days on a 12 metre raft with a crew of eight men. The only way he was able to resist was to reason that 'if she were really there, I would have noticed her before.' Yet for him, she was as real as the thirst and the heat and the isolation. Where had she come from? Norman may have wanted to believe she had swum aboard in mid-ocean, but he wasn't irrational, at least not yet. She hadn't come from the sea; she had come from his mind. False impressions can be as convincing as real ones. Enclosed in a protective skull, the brain cannot see, hear, smell, taste or feel. Just as a military commander inside a bunker must rely on reports from his soldiers in the field, the brain depends on incoming electrical and chemical signals to piece together a picture of what's going on outside. If the information received is false, the picture will be false. The only way to know that, though, is for the commander, or brain, to apply logic and deduce that the picture painted by the information is impossible. The battle to survive requires your brain to continually interpret incoming signals for danger or opportunity. Your brain remains sensitive even while you sleep, albeit at reduced vigilance. It is designed for constant activity. If the level of outside stimulation falls too low, it will pick up and intensify signals from within, or even make them up. A phenomenon called 'phantom limb pain' sometimes occurs after an amputation when the brain, faced with a sudden complete loss of incoming signals from the lost body part, begins to generate the same signals as if the part were still there. I once treated a bass player after a drunken patron at a New York jazz club took exception to his music and shot his hand off. The musician said that for months afterward, whenever he heard a bass being well played, he felt his missing hand 'come alive with pain.'

If the brain doesn't get enough excitement, it will create its own. A lot of electrochemistry is generated by emotion and memory. The signals they send are powerful and highly organized. The visual cortex, the area of the brain that monitors input from the eyes, will light up on a brain scan almost identically in a person who is imagining a scene as in a person who is actually seeing it. Usually the false image is less intense because it's mixed with outside signals the brain is receiving at the same time, but if those signals are absent, there's no competition, and emotion and memory have free reign to conjure up any image they want. In the case of Norman Baker, mindlessly splicing rope under a hot sun on a long voyage with an all-male crew, it's easy to understand why he saw a pretty waitress offering him a drink. His hallucination was all the more vivid because he wanted so much to believe it. It was only by using his highest capacities of reasoning and discipline that he was able to overpower the intense signal and make the image disappear – to his great disappointment.

Norman Baker set himself adrift intentionally, and so did Frenchman Alain Bombard, a voluntary castaway who in 1951 crossed the Atlantic in a rubber raft without supplies of food or water. He set out from the Canary Islands and drank nothing but seawater during his first seven days adrift. Once he began catching fish, he ate and drank them, alternating the fluid he extracted with small amounts of seawater. He landed in Barbados sixty-five days later, proving dramatically that man could live off the sea indefinitely.

Nonetheless, Bombard had help. Besides the simple fishing gear that most life rafts carry, he had books and musical scores, the morale value of which would be hard to calculate. He brought other possessions too: a comprehensive knowledge of sea life and the confidence that it could and would sustain him. And he had one additional asset that castaways rarely have: the resolve of someone who has chosen his course voluntarily and has something to prove to the world.

Unlike Bombard or Baker, most people who find themselves adrift at sea have no prior intention of starting an adventure. Someone who sets off in a life raft with the determination to cross the Atlantic – and read some good books – enters the ocean with a very different mind-set than someone expecting a 32 km island-hopping ferry ride in the Seychelles – an archipelago in the Indian Ocean 1,609 km off the coast of Africa. The ferry was the *Mary-Jeanne*, an old scow powered by a car engine that, shortly after it set off, got stuck in low gear, consuming a lot of fuel and making very little headway. The boat ran out of gas in sight of the destination. The crew dropped anchor and waited for help. That night there was a storm, the anchor chain broke and the boat drifted out to sea. The ten passengers and crew, mentally prepared only for a two-hour boat ride, faced what would become a seventy-four-day ordeal.

Expecting an early rescue, they quickly used up their meagre supplies of food and water, then passively waited thirteen days before trying to catch a fish. After one attempt with a bent wire and one attempt with a makeshift harpoon, they gave up. Periodic rain quenched their thirst, but they were dying of hunger.

Hungry means that not enough fuel is being distributed by the blood. Blood transports fuel in the form of a sugar called glucose, a simple carbohydrate that enters cells easily and burns quickly (this is why eating sugar gives you a quick burst of energy). Blood resupplies its load of glucose constantly by tapping into the liver and muscles for their stores of glycogen, a slightly more complex carbohydrate readily converted to glucose as needed. Glycogen reserves are normally maintained at about a three-day inventory, with regular restocking from outside sources through a process called eating. When food delivery stops and carbohydrate supplies start running low, the blood goes back to its main warehouse – fat cells. Fat does not burn as cleanly as carbohydrates. It leaves behind a residue of acetone, excreted in urine and exhaled in breath, making starving people smell like nail polish remover.

With the steady mobilization of fats in tandem with carbohydrates, and no incoming food to replace the dwindling fuel supply, the body takes steps to decrease the rate of power consumption. It turns down its idling speed, or basal metabolic rate, and shuts down nonessential functions. The result is like a brownout. There is a slowing down of reaction time, decreased production of standby cells that repair wounds and fight infections and diminished formation of 'superfluous' chemicals such as sex hormones, since reproduction is not exactly a priority during starvation. This conservation of energy takes place within about two weeks, which explains why dieters often find it increasingly difficult to lose weight after a promising start.

The continuing energy crisis leads to the exploration and development of alternate energy sources. The body has vast reserves of protein, which comprises over one-third of body weight, though none of it is stored as fuel. Proteins provide structure and function to every system of the body, including muscles, bones, liver and kidneys. They can be burned as fuel, yielding about as much energy as carbohydrates, but it's a lot like burning your house down to keep warm.

Although burning protein is a sign of the body's desperation, it is nonetheless done in an orderly manner. Taken together, muscles are the largest organ in the body, and they are made largely of protein. They account for twice as much body weight as fat. Because fat burns twice as efficiently, however, muscles represent about the same amount of potential energy as fat. Muscle is the first protein to be sacrificed, because it is abundant and the loss of even a lot of it will not have an immediate impact on survival; weakened muscles can still function. More critical are the much smaller quantities of protein that make up tissues in the liver, brain, heart and kidneys.

So the starving body maintains a rigid hierarchy when it accesses its own fuel sources. After carbohydrates start to run out, it taps fat supplies, then muscle reserves, until all that remains are the proteins in the critical organs and a tiny amount of fat essential both as a key ingredient in hormones and as electrical insulation for nerve and brain cell transmissions. At this point, body weight has been reduced by 50 per cent. The body is eating itself alive.

After thirty-six days at sea, the passengers and crew of the *Mary-Jeanne* were living in decomposing bodies. Their only external supply of nutrients had been a few birds and some flying fish, which were quickly devoured when they landed on the deck. At one point they had gone eighteen days with no food at all. On the thirty-sixth day they sighted land, drifting close enough to see houses and trees before a shift in current carried them back to an empty horizon. The passengers lost hope and, with it, the last remnants of their strength. Their bodies were no longer able to 'defend' their essential fat and protein. And once those are consumed, body systems go haywire; chaos and death ensue.

So it was for those on board the *Mary-Jeanne*. Passengers started to die. The bodies thrown overboard were eaten by sharks. Still, it never occurred to the survivors to try fishing again, even if it meant using human body parts as bait. They were starving to death with food all around them because they were mentally unprepared for their hardship and were unable to generate any will to survive. On the seventy-fourth day, the boat was finally spotted by an Italian tanker, and two living corpses were rescued.

Perhaps more passengers would have survived the ordeal had their options not been limited by one of the strongest of cultural taboos: cannibalism. Although they were starving, their civilized upbringing had created within them an artificial barrier so powerful and unquestioned that it was able to block their access to a readily available source of nutrition. They viewed their dead shipmates as departed souls but not also as food. An inbred horror of cannibalism works against the individual but favours the survival of the group by preventing its members from becoming targets for each other. Yet the revulsion is not universal. In most societies, only the most deviant members would eat another human, but there are others in which, at least until recent times, the practise was accepted. Some even revered it as a way to capture a victim's soul. Whether for the spirit or for the nutrition, there is no biological reason why one human cannot eat another. Indeed, the meal would be more balanced than turtle or fish. Unlike seafood, humans contain carbohydrates, and because the muscle has a much higher fat content, the meat would provide more energy. Our liver, heart and kidneys are especially nutritious and would supply essential vitamins. Given that the protein intake would already be in human form, it could be utilized efficiently; there would be no need to rearrange many of the amino acid building blocks. Dressed out like an animal, a human can yield as much as 27 kg of meat – about twice as much as a sea turtle. However, only the most hardened survivor would find all of it to be edible. Everyone else would have to remove the head, hands, feet and genitals – in other words, all the most obvious signs of human identity. I am often reminded how personally we identify with that anatomy when I see medical and nursing students studying gross anatomy or watching surgery. Many will faint or ask to leave the room when those body parts are being dissected or repaired. Conversely, they are the parts a crazed warrior would be most likely to remove from his vanquished enemy as a sign of complete victory.

For a starving castaway, cutting up and eating a dead human body is not an act of perversion. It is a demonstration that a taboo imposed by civilization no longer prevails; it yielded to extreme stress and hunger. Such was the case in the Andes in 1972 when a plane crash turned a Uruguayan rugby team into a group of desperate survivors surrounded by frozen corpses. They were stranded for two months in barren, frozen terrain with no source of food outside their wrecked airplane. The players, after much soul-searching, brought themselves to eat their former teammates, remaining strong enough to make forays to lower elevations, which eventually led to their discovery and rescue. They were saved because they had committed an act of self-preservation consistent with the prohibition against suicide set forth by the Christian Bible with which they had been raised.

Breaking a taboo is dangerous because the proscribed behaviour is then no longer enforced by fear, only by ethics. Once broken, an individual might be tempted to bring down other cultural restraints in order to minimize guilt over his irreversible transgression. Concepts such as fairness and morality, which reside in the cerebral cortex, can easily be deactivated by self-serving rationalizations. As ethical standards drop, the cortex exerts less control over primitive urges, and the human being will steadily fall more under the command of his most primal instincts.

This degradation was evident in the behaviour of some members of the Donner Party, settlers in an 1847 wagon train bound for California but marooned by an early winter in an icy mountain pass of the Sierra Nevada. The first settlers to die were properly buried. Later, however, as the food ran out, they were dug up and eaten. When the situation became even more desperate, those who died were roasted and eaten immediately. The acts seemed to unleash primal behaviour in some of the members, who soon found themselves able to kill and eat the weakest among them, reasoning that they wouldn't survive anyway. Applying their own moral distinctions, they then found it acceptable to kill any of the Indians travelling with them, since they were of an inferior race. One man may have resorted to killing and eating a woman settler, although none had the idea of killing members of his own family, a taboo that runs so deep it was God's ultimate test for Abraham.

Even in those societies where killing and eating a human is countenanced, it is never done casually. The practise is generally accompanied by ritual and solemnity – whether the victim is providing strength for a triumphant warrior, magic powers for a medicine man or nourishment for a starving sailor. The tradition of cannibalism once travelled the high seas in the mind of every mariner on every voyage. When a crewman signed on, he entered a culture in which rare and dire circumstances might make cannibalism an acceptable survival technique. All knew that a shipwreck could turn a crew into a group of hungry humans eyeing each other as food. The rules were unwritten but clear, and following them made the deed seem less savage. A dying sailor would be sacrificed first. There was no point in having him consume precious supplies, and the longer he lingered the more he would dry out, making him less nutritious and harder to chew. It was believed that drinking the blood while it still flowed would quench thirst. Actually, this isn't true, since a dying sailor's blood would most likely have an even higher salt content than that of his slightly healthier shipmates. The macabre soup would have a desiccating effect similar to seawater.

Surviving sailors, regardless of rank, could also be sacrificed provided the selection was done fairly, such as by drawing straws or some similar method. The first round would decide who the meal would be, the second who the butcher would be. Sailors would be expected to accept their fate as their duty. The spectre that one day they would have to play out that drama must have haunted many a ship. To defend against a fear they are powerless to control, humans often invoke gallows humour. Something cannot be as horrible as imagined if it can be trivialized. Thus 'The Yarn of the *Nancy Bell*', about a ship's crew with no food or water: 'For a month we'd neither wittles nor drink, / Till a-hungry we did feel, / So we drawed a lot, and accordin' shot / The captain for our meal.'

Humour can play a part in trivializing the deed as well as the fear. In 1884 four crewmen were adrift in a dinghy for three weeks after their 15-metre yacht *Mignonette*, sailing from England, sank in a South Atlantic storm. One of the crew drank seawater and became delirious. The others decided he couldn't survive, made a sham of drawing straws, then killed and ate him. Three days later they were picked up off the coast of Brazil. The sailor who had been the butcher related later that they were rescued 'as they were having breakfast, with their hearts in their mouths.'

The other extreme is to revert to a feral state, the way pets are forced to do when turned loose to survive on their own in the wild. In 1821 a vengeful whale rammed and sank the whaling ship *Essex* in the Pacific. The nearest land, the Marquesas, was 1,931 km away, but the crew, twenty souls in all, were afraid to sail their lifeboats there because they feared that cannibals inhabited the islands. Instead they attempted a 4,828 km journey to South America. Three months later, eight wild-eyed survivors were found in boats with human remains scattered about. They were jealously clutching human bones and recoiled from their rescuers, afraid that their food was being taken away. They had become exactly the cannibals they had feared.

Not all castaways become cannibals, but none can be fussy eaters. Staying alive means adapting to the realities of their new environment. And these govern more than just food. Sailing ships are mobile outposts of civilization, and when they sink, they strand their inhabitants in an elemental world of sea and sky. Suddenly bereft of the support structures on which they have leaned, their fate depends on whatever knowledge and experience they can draw on and what resolve they can generate. The animals now surrounding them have had millions of years to perfect their means of survival. They have stronger muscles and faster reflexes, but humans have bigger brains – they can outsmart them. To do that, these people have to become inhabitants of the sea, focusing all their abilities on simple goals: satisfying thirst and hunger, avoiding pain, overcoming fear. There is no recognition of their achievements beyond survival. Fate depends on will. The initial impulse to wait passively for rescue, or even death, must be conquered. Strength can coalesce around defiance, as it did for Dougal Robertson when the freighter passed him by, or it can emerge, steady and steadfast, from deeply held religious beliefs, as it did for his wife, Lynn.

Marilyn Bailey didn't consider herself a religious person, but she believed that a supernatural power was governing her affairs. When two ships passed the Baileys by in two days, Maurice began to lose hope, but Marilyn understood why they had made it so far. Their time together in solitude had given them a chance to begin again. She had planned out a new life for them, and firmly believed she would live to carry it out.

Steve Callahan drifted alone in a rubber raft for seventy-six days – two days longer than the *Mary-Jeanne* had been adrift – yet he survived, while eight of the ten passengers on that ferryboat did not. Callahan had the one thing they lacked: the determination not to let fate overcome him. He divided his 'self' into three parts: the physical self that feels pain, the emotional self that feels fear and the rational self that takes control over them both. At first his responses to pain and fear were instinctive reactions that protected his higher functions. As his ordeal wore on, though, more and more he needed those higher functions – of reasoning and will – to maintain control over his once automatic responses. It became harder and harder for him to 'coerce' – his word – his body and his emotions. He was gradually losing his 'ability to command.' He knew that if that was lost, he was lost.

What all survivors have in common is an energy flow from the top down. Whether the source is defiance, religion, positive thinking or willpower, it generates a spark in the cerebral cortex felt as motivation. A cascade of electrical and chemical brain circuits is activated to coordinate the thinking and behaviour that converts a passive human into a streamlined survival machine.

The sailors on the *Mignonette* and the *Essex* who killed and ate their shipmates were also survivors, but their energy was feral, primitive. Hunger and fear are powerful motivators, exciting the amygdala and the hippocampus. These 'lower' brain centres generate compulsive activity, which in a human can be dampened by cortical control. Within the cerebral cortex are standing circuits, developed in individuals to varying degrees, which, in a very simplified sense, are the physical manifestations of morality and reason. These electrochemical circuits form counteracting signals – experienced as inhibition – that impede or neutralize upwelling electrical and chemical power generated by primitive drives and emotions. This is why someone with cortical brain damage, such as after a fractured skull, might undergo a personality change, becoming more belligerent, for example. Pushed to the extreme, a castaway might also have his cortical inhibitors overpowered by stress, or uncoupled by rationalization. Either way, unleashing of primitive responses promotes his own survival but sacrifices the essential quality that makes him a human.

Rafts that stay afloat eventually reach the shore or another boat. Steve Callahan crossed the entirety of the Atlantic, floating from offshore Africa to the Caribbean island of Guadeloupe, where he was picked up by a small fishing boat. After 117 days, Marilyn and Maurice Bailey were brought aboard a merchant ship bound for Korea. The Robertsons, with the four boys, were spotted by a Japanese freighter off the coast of South America after thirty-eight days adrift. Dougal said, with typical British understatement, that the 'rescue came as a welcome interruption of the survival voyage.'

With the advent of global positioning satellite receivers, satellite phones and water-activated emergency signalling beacons, one might think that shipwreck odysseys would be a thing of the past. One would be wrong. Ask Richard Van Pham or Terry Watson. Van Pham was rescued by a US warship off the coast of Costa Rica after drifting 4,023 km in the Pacific for nearly four months. He survived by collecting rainwater in his sail, catching fish and bashing turtles and seabirds with a bat. Van Pham's 8 metre sailboat, *SeaBreeze*, had become disabled when a sudden storm broke its mast, outboard motor and radio after he set out on what he thought would be a 48-km trip from Long Beach, California, to Catalina Island in the summer of 2002. Less than three weeks after Van Pham's rescue, a fishing boat off the coast of South Carolina spotted Terry Watson adrift on his broken-masted 7 metre sailboat, *Psedorca*. He had been reported missing nearly three months earlier, shortly after he left Miami for a trip to Bermuda. Despite a massive search, covering more than 20,720 square kilometres (an area larger than Massachusetts) the Coast Guard had been unable to detect any trace of him. When finally found, Watson was unable to explain what had happened to him. He was emaciated delusional and, at first, unwilling to leave his boat. To a castaway abruptly plucked from his adopted environment, civilization can initially be more frightening and hazardous than the sea. Along with a dizzying barrage of people, noises, smells and images, abundant food and drink suddenly become available. Having lived on too little fresh water and too much salt air, a shipwrecked survivor will most likely have blood that is reduced in volume and very salty. Then, when large quantities of fluid enter the body rapidly, the blood will draw water in to dilute itself back to a normal concentration and volume. The rapid rise in fluid load may prove too much to pump for a heart unused to exercise and perhaps even structurally weakened by having lost some of its protein to the metabolic fire. Kidneys ordinarily filter out excess water, but they have probably not yet unclogged from all the salt they endured during the voyage. Fluid will back up and collect under the skin, causing a condition called oedema, bloating the face and body – and especially the legs, since water tends to sink.

Food intake must be gradual. Digestive muscles will be too weak to move large quantities of food if they need that food to rebuild themselves and the other organs. The first meals must be small, frequent and easy to digest, balancing carbohydrates and fats for energy with proteins for structural repair. Like water, too much too soon can be fatal, but usually it is just uncomfortable. Indigestion and diarrhoea are signs that the body has not yet switched out of survival mode.

Being shipwrecked is a very effective method of dieting. Steve Callahan, Marilyn and Maurice Bailey, and Lynn and Dougal Robertson, none of whom were fat to begin with, lost a total of almost 91 kg, reducing their weight by nearly one-third. Since death ensues (in non-obese individuals) when body weight drops by about one-half, they were riding the seas with a very thin safety margin.

When setting out on any long ocean voyage – particularly a hazardous one – it is advantageous both to be overweight and to have an efficient metabolism that produces fat easily. Individuals with a large internal energy supply have a better chance of survival. Besides being better suited to open boat travel, they are better able to colonize a new land since it might take several months after arrival to grow and harvest an adequate food supply. In an environment consisting of many small islands separated by large expanses of ocean, natural selection favours people who put on weight easily and lose fat slowly. The part of the world with that environment is the South Pacific, and the branch of humanity that successfully expanded into it is the Polynesians. They are the closest humans have come to being inhabitants of the sea.

Not having the navigational capabilities of migratory birds or oceangoing fish, the Polynesians were nevertheless able to create an ecological niche for themselves by making astute use of the senses they did have. Crossing empty horizons, they learned to locate land by indirect clues. A greenish tint on the undersurface of a cloud was light reflecting island vegetation. Birds that roost on land could be followed home in the evening. Islands generate sounds when breakers crash against the shore and emit smells when wind carries the scents of flowers, fruits, and earth. Waves that strike an island bounce back and radiate outward, creating a ripple pattern that can be seen on the surface and felt against the boat.

Polynesians adapted to their environment primarily with their most valuable and adaptable organ, their brains. But their bodies adapted as well, gradually evolving a greater ability to make and store fat. Navigational skills and metabolic efficiency were critical for survival in the South Pacific until the area was overtaken by technology. Now, with electronic navigation and readily available food, the rules for survival have changed. Polynesian minds can quickly learn to use global positioning satellite receivers and radar, but their bodies are not that nimble. Their specialized adaptations for risky sea voyages have made them especially vulnerable to the dangers of overeating. The population has far higher than average rates of high blood pressure, diabetes and massive obesity. Polynesian women suffer more than men. Women in general seem to retain fat more easily, and while this may lead them to fret over their appearance, it makes them less prone to starvation. Relatively more of their fat is under the skin. This gives them softer curves, but it also provides an extra quantity of reserve fuel. Women also have less body mass to support and require less energy. Furthermore, in survival situations, men are more likely to be doing the physically demanding work, thereby burning up more calories. Some or all of these factors may explain why Marilyn Bailey was in much better shape than Maurice when they were rescued, and why the women in the Donner Party fared better than their husbands.

Damage done to a body by lack of food and water is repaired only slowly. Spindly legs have to be rebuilt and reconditioned. They are drastically affected because they are the largest source of muscle protein, and what wasn't robbed from them wasted away anyway. To stay healthy, muscles must be used, and on a life raft, there aren't many occasions to go for a long walk, or even to stand. With fats and proteins no longer being diverted for fuel, there are enough around to restart dormant body functions and repair lingering wounds. Hormones are synthesized again, giving Marilyn Bailey her first menstrual period in three months. Skin ulcers that had caused Maurice Bailey so much pain for so long suddenly healed.

The reminders of their life on the ocean receded. The castaways were no longer preoccupied by thirst and hunger or haunted by the certain knowledge that they could not hold out indefinitely against the power of the sea. Released from the imperatives of survival, they experienced the unfamiliar sensations of security, warmth and comfort. They left the raft and rejoined society, rediscovering the luxury of moving around freely, with a delightfully unyielding surface beneath their feet. Life returned to what is normal for humans shielded by a protective civilization, where survival laws are not strictly enforced. Success no longer means merely staying alive; that's too easy when food and shelter are readily available even to those who are not strong or smart or determined. Goals become more abstract: the acquisition of respect, recognition, approval, and then beyond that, the creation of ideas, art and beauty. To paraphrase John Adams, 'I study the strategy and tactics of war so my grandchildren can study music.'

Shipwreck survivors have won their battle with the ocean, and their prize is the chance to resume their more protected lives. The abilities and the intensity that they called forth to allow them to stay alive on the high seas are largely withdrawn – superfluous in a society where the survival instinct is not critical to survival. They resume or take on roles as fathers, mothers and friends, and work as anything from sailors to storekeepers. In short, they become nearly indistinguishable from the rest of the population who were never put to the test. They have proven they have the qualities for survival, but most of us will never have the chance to look for those qualities in ourselves.

For a survival epic to become a story, someone has to live to tell it. People can be lucky, like the French couple in the Mediterranean, or unlucky, like some of the students on board the *Albatross*. They can survive on sheer will like Poon Lim or on raw animal instincts like the whalers on the *Essex*. But the harsher the conditions and the longer the isolation, the stricter the selection becomes. The circle of survivors tightens and then disappears as bodies and minds are pushed to their limits and beyond. We only know the stories of the winners. The greatest battles against the sea are likely to be the ones that ended in defeat, and thus their stories will never be told.

DESERT

THE MARATHON OF THE SANDS

ON THE FOURTH DAY OF AN ULTRA-MARATHON through the Sahara Desert, Mauro Prosperi got lost. He survived a sandstorm, then with only a finger's breadth of water in his canteen, crossed 209 km of dunes over nine days, in temperatures above 38°C, before being rescued by an eight-year-old Tuareg girl. He is either the most incredible example of desert survival ever or the most elaborate fraud in the history of endurance sports. Can a human body, even one belonging to a superbly trained athlete, survive nine days in an oven?

The Marathon des Sables is an annual 257 km race through the dry ocean of the Moroccan desert. Competitors are immersed in superheated air that flows over their skin and fills their lungs, bathing them in heat outside and in. They run over waves of sand – a thick, dry fluid that splashes up as each footstep sinks below the surface. Overhead is a nuclear reactor, the sun, that beams radiation through an atmosphere too thin and transparent to protect the trespassers crossing underneath.

There were 137 trespassers in the 1994 race, each one carrying his own portable micro-environment to defend his body against the surroundings. Except for water stations at the checkpoints, the racers traversed the desert self-contained, with food, spare clothing, a sleeping bag and emergency supplies in their backpacks. They were competing against the other athletes but, even more, they were competing against the desert.

Deserts form in those parts of the earth that receive less than

254 mm of rainfall a year. About one-fifth of the world's land surface qualifies, usually because it lies on the far side of a mountain range tall enough to create a barrier to the moisture-laden air that collects over oceans. The Sahara is the largest desert in the world, covering half of Africa, with a total area roughly equal in size to the continental United States. The Atlas Mountains along Africa's western coast block airflow from the Atlantic. Ocean winds push the air against the mountains, and as it rises, the air cools. Because cold air cannot hold as much water as warm air, the water drips out as rain or snow, so that by the time the air has made it across the peaks and slid down the other side it has effectively been dried out. With no clouds and no moisture to absorb, deflect or diffuse the sun's rays, solar energy strikes the ground at full intensity, overheating the air and baking the land into sand.

Wind is created when heated surface air rises and surrounding air moves sideways along the ground to replace it. Most deserts, being open expanses, allow the wind to travel great distances in one direction at a steady speed, rolling sand along to form dunes, which can sometimes grow to be 305 metres high and 24 km long. Coaxed along by the wind, dunes advance in orderly patterns, like waves. But where air heats up unevenly, hot pockets form that rise suddenly, like hot-air balloons. The strong wind sweeps up and around these pockets, pulling sand grains into its vortex, and the combination becomes a sandstorm.

At 32 km into that day's 80-km run – the fourth and longest leg of the marathon – Mauro Prosperi was maintaining a good pace, so good that by early afternoon he was in seventh place. But he was about to face a much tougher array of opponents. The temperature had reached 46°C. The superheated air was stirring up swirling winds. Suddenly, Prosperi was enveloped by a sandstorm so violent that, unbeknownst to him, race officials suspended the day's run. Despite the incredibly poor visibility, Prosperi thought he could still see the trail. He kept on running.

Windblown grains of sand pierced his skin like needles, causing his nose to bleed and cutting the inside of his throat. His competitive spirit could not keep up with the increasing ferocity of the storm. Finally he stopped, crawled into a bush and wrapped a towel around his face. All day the wind whipped the dunes into a raging ocean, forcing Prosperi to change locations several times lest he be swamped by a wave of sand. Eventually, night fell. By morning the sandstorm was over.

When Prosperi opened his eyes he saw only sand in every direction. What had been a competition among athletes was now a contest against nature. His goal was no longer to win the race, but to stay alive.

The Sahara desert is not obviously compatible with human life. Daytime temperatures routinely soar above 38°C; night-time temperatures can fall well below freezing. To the untrained eye, food is scarce, water nonexistent. Few features rise above the sand, offering little chance for shelter and little aid to navigation. The Tuareg people of western and central Sahara, through evolutionary adaptation and generations of accumulated wisdom, manage to survive here, scattered as nomads wringing subsistence from the desert. But Prosperi was a crowd-control policeman from Italy, and this was his first time in the desert.

It was also the first time in the desert for the Hughes family. On vacation from England, Andrew and Jane Hughes and their preadolescent sons, Matthew and Sam, on their trip to Tunisia in 1989, became bored hanging around the hotel pool. They decided to rent a car and drive to the southern market town of Duse, at the edge of the Sahara. Thanks to bad maps and few signposts along a barely discernible road, they were soon lost and stuck in the sand. Believing they were close to Duse they got out of the car to go the rest of the way on foot. After an hour of walking with no town in sight, Andrew told Jane and the boys to return to the car and wait. He would get help and come back for them. Jane and her sons set off under a ferocious sun, carrying a litre and a half of water. They weren't worried; earlier they had passed some big water tanks. Andrew walked on by himself.

Humans are designed to endure heat far better than cold, but no one can withstand prolonged exposure to a blazing sun and no one can live long without water. The desert, however, offers ample opportunity for lost travellers to try to survive both. Humans must fiercely protect their internal temperature, for it holds the key to all their life functions. The human body is a mass of millions of exquisitely sequenced chemical reactions, which speed up as temperature rises. Individual changes in the cadence of those reactions will quickly lead to internal chaos, like a symphony orchestra with each member playing at a different tempo. The timing, and thus the temperature, of these reactions is so critical that if body temperature varies by more than 2.3°C from 37°C, systems begin to malfunction and the body's formidable defences start to crumble.

Years ago, when survival was first being examined from a scientific point of view, it was assumed, reasonably enough, that people who had adapted to desert life would have higher body temperatures than those adapted to places like the Arctic. In fact, regardless of where we live or for how many generations, all people guard the same internal temperature. We have indeed adapted to our respective environments, but we have done it by modifying our body systems and our behaviour.

Heat enters or leaves an object in three ways. Where there is direct contact between two objects, heat flows from the warmer one to the cooler one – such as from sand dunes to hiking boots or from hot air to clothes – via conduction. Convection, the second form, is a kind of facilitated conduction that occurs when air circulates over one of the contact surfaces, spreading the heat faster. This is why a heater with a fan makes a room warmer faster than an oven with its door open – and why desert winds very quickly increase the temperature of human skin. The third form of heat transfer is radiation. What we experience as heat is actually the vibration of molecules. The faster something vibrates, the higher its temperature and the more energy it gives off. The sun, whose surface temperature is about 6,093°C, releases vast amounts of energy in the form of waves, which radiate through space until they collide with something, such as a human being. This in turn speeds up the molecular vibrations within that human. These sped-up vibrations are what we sense as 'getting hotter'. In the desert, direct solar energy accounts for about two-thirds of the heat load absorbed by the body.

Proteins, whether contained within eggshells or skulls, congeal at temperatures above 44°C; desert temperatures routinely surpass 49°C (temperatures in saunas can rise to over 82°C), yet brains don't become hard-boiled at these temperatures. And heat doesn't come only from the outside. Because no biochemical reactions are 100 per cent efficient, they all give off some heat as a by-product. The busiest organs, the brain, heart, lungs, liver and muscles, generate the most. The heat is transferred to the blood (by conduction) and then circulated through the body (convection) to maintain what we call body temperature.

In temperate climates, air temperature is generally lower than body temperature, and as a result, body heat is constantly being given off into the space around it (radiation). The rate of heat loss depends upon the temperature differential between the environment and the body. When the air is 9°C cooler than the body, the rate of heat production is exactly offset by the rate of heat loss. This means that the human body will be in optimum heat balance when the outside temperature is 28°C. That's the average temperature on the African plains; one solid piece of evidence that human life evolved there.

The temperature balance of 28°C applies only to a body at rest, emitting the baseline heat level called the basal metabolic rate. As soon as we start exercising, the metabolic rate – and consequent heat production – increases enormously. Unless the heat can be quickly dissipated, one hour of intensive exercise will raise body temperature to 60°C. Even on the temperate African plains, a man who evades a lion or captures an antelope still requires an effective way to rid his system of excess heat if he is to survive there.

So too do Italian marathon runners and British tourists lost in the desert. Temperature regulation is critically important for all humans, yet, strangely, we have no systems designed specifically to cool our bodies down, other than sweat glands, which are actually highly modified hair follicles. Thermostatic control depends on organs and tissues from other systems. The body recruits blood vessels, skin, fat, muscles and most important of all, because of its ability to modify conscious behaviour, the brain.

To coordinate an effective response, the body must first gauge the outside temperature, so that it can begin to respond long before its core temperature becomes affected. The entire outer surface of the body is supplied with nerve endings called thermoreceptors, thermometers sensitive either to hot or to cold. Heat receptors fire more frequently as temperature rises; cold receptors fire more when temperature falls. The receptors are fine-tuned to 28°C – the optimal outside temperature for body chemistry.

Signals from the thermoreceptors are transmitted to the hypothalamus, the brain's maintenance centre, located at the base of the skull. The front of the hypothalamus contains the body's thermostat, actively monitoring internal body temperature while also remaining exquisitely sensitive to skin temperature, its vital early-warning system. When the hypothalamus starts receiving increased signals from the heat receptors, it takes control of blood vessels and sweat glands and adjusts their function to facilitate cooling. Without the alarm set off by changes in skin temperature, the hypothalamus would be unable to react to an increase in core body temperature until after it occurred. This would be highly dangerous, given that a 2.3°C differential is enough to disrupt the body's functions. Should the thermostat in the hypothalamus itself become disrupted, the entire thermoregulatory system would rapidly spin out of control.

Our body's margin of survival is precariously thin. We spend our entire lives less than 5.5°C away from fatal overheating, a frightening thought on a planet where temperatures can vary by more than 50°C. Clearly, our bodies need a reliable cooling system.

Not being too furry, we humans lose most of our heat through our skin, which acts like a car radiator. Blood, the liquid coolant, passes through all the organs, picking up heat and carrying it to the skin surface, where it is cooled by proximity to the outside temperature. Blood is distributed from the heart to the organs via a system of tubes. The big conduits are called arteries, the finer ones, arterioles. These tubes have muscles within their walls that regulate the amount and direction of blood flow, distributing some under the body's insulating fat layer and some to the skin surface, where it is cooled. The ring-shaped muscles can contract to narrow the tubes or relax to dilate them. They respond to signals from the hypothalamus to divert more blood to the skin whenever the body starts to overheat. The system is fine-tuned and elegant, but by itself not enough to keep us out of the danger zone. For the blood to be cooled by air, the outside temperature must be significantly below that of the skin, whose temperature is 35°C. Exposure to higher environmental temperatures, such as routinely occurs in the desert, actually heats the blood. Moreover, muscular activity greatly increases the internal heat load that must be dissipated through the skin. Another system, no matter how inelegant, is needed for thermal regulation. That system is sweating.

Extruding water from the body may be disagreeable in social situations, but it works. Sweat glands spread a layer of water on to the surface of the skin. As the water evaporates – a cooling process – it draws heat from the body, vaporizing it into the air. Sweating can dissipate heat twenty times faster than blood-cooling. When tinkering with blood vessels isn't getting the job done and the skin is heating up, the hypothalamus turns on the sprinkler system.

About 3 million sweat glands are distributed unevenly on the skin. Most are located on the forehead, face and neck, chest and back, and in the armpits and groin – areas that benefit the most from cooling because of the large volume of blood circulating just beneath the skin. For sweating to be effective, however, the water has to evaporate. High humidity is so uncomfortable because the air, already filled with water, has little capacity to add more, so sweat simply sits on the skin's surface.

When air temperature approaches body temperature, you need to sweat to stay alive. The higher temperature speeds up evaporation and actually improves cooling efficiency, which would otherwise steadily diminish and then reverse as the outside temperature rose. Sweating may be essential for survival in the heat, but in the desert, unless you have enough water, sweating will kill you.

Two-thirds of the human body is composed of water. The average person contains about 50 litres of fluid and loses a minimum of 2 litres in daily body maintenance. The kidneys use water to flush out waste as urine; the lungs use it to moisturize inhaled air so that it won't irritate the sensitive pulmonary linings. Some also seeps out passively through the skin – the body's not-quite-watertight container. When the body turns on its sweat glands, however, water losses mount rapidly. The hypothalamus controls both the number of glands activated and the rate at which the sweat pours out. Sweating is a profligate waste of water, but the body knows no limits when it's fighting an internal fire; at rest on a hot day, it can easily use up 5 litres. When the body is down even 1 litre, its function becomes impaired. Once it is down 5 litres, fatigue and dizziness set in. A loss of 10 litres disturbs vision and hearing and sets off convulsions. A deficit of 15 to 20 litres, roughly a third of the body's total amount of water, is fatal.

To make matters worse, exercise and anxiety greatly accelerate sweating. Walking can use up an additional litre an hour, and sweating is also the automatic response to being nervous. So in the desert, the body stresses of heat, exercise and anxiety all contribute to create maximum water loss. Surviving in the desert resembles trying to control a fire without an external water supply. Should the body's core temperature rise by 1°C, the speed of metabolic reactions increases 15 per cent. At 41°C, the speed will be about 50 per cent above normal. Beyond that temperature, reactions accelerate even faster, and at 43°C, the brain is cooked.

To sustain the increased blood flow and the production of sweat, the body needs to resupply itself with water. Finding it requires conscious action. The hypothalamus sends a water-seeking signal to the cerebral cortex – which we experience as the sensation of thirst. The thirst signal does not get triggered immediately. Sensors in blood vessels are continually 'tasting' the salt concentration of the blood. It takes a loss of about 3 per cent of body water before blood becomes salty enough to set off the alarm (the alarm will also go off if excess salt is ingested through foods). There is practical value to the delay, since it means that humans are not slaves to small changes in salt concentration. We are free to perform other activities if we are not driven to drink constantly.

The body can tolerate a fluid loss of about 5 per cent before developing any obvious symptoms of dehydration, such as dizziness or fatigue, but even a 1 per cent loss can impair normal functioning. Given the prevalence in our diets of salty foods and dehydrating drinks such as coffee and tea, most of us are chronically underhydrated – at a level that triggers no alarms but nonetheless subtly affects our performance. When we drink enough to bring the deficit to within 3 per cent of body water, our thirst is quenched. The problem is easily solved – except when there's no water to be found. Lacking an adequate supply of water, the blood remains too concentrated and the body must therefore draw down its own reserves to dilute it. Water is distributed in the blood, where it flows freely, and in the organs, where it is held like a wet sponge. To manage thirst, the body shifts water to the blood in the process of osmosis, wringing it out of the organs until the salt concentration in the blood returns to normal. Osmosis has its own limits. It will not allow so much to be taken from the organs that they become even saltier than the blood, but will keep the two concentrations at equal levels. This means that both blood and tissues will dry out and become salty at the same rate, with water continuing to seep out of the organs. The body will conserve water where it can. The kidneys clamp down on outflow, flushing out waste using less and less water. A progressively darkening colour and a decreasing amount of urine are far more accurate indicators of dehydration than is thirst.

With no water coming in, the body will make do with what it has. It will set priorities. In the desert, the body must dissipate heat, continue oxygen intake and develop a survival strategy. Therefore it will shunt blood to the skin, the lungs and the brain at the expense of other less immediately essential organs, such as the stomach, liver, intestines and muscles. Faced with the need to supply more blood to the network of heat-dissipating vessels that have opened up under the skin, while still maintaining flow to the brain and lungs, the heart pumps harder and faster. The chest pounds and the pulse rises. To aggravate matters, the drier blood is thicker and therefore harder to propel through the vessels. Increased work generates more body heat, which requires even more blood flow to the skin, creating a vicious cycle. Blood flow can't keep up. Eventually even the most vital organs will become deprived and the body will collapse into heat exhaustion.

Should he happen to fall in direct sunlight, an exhausted desert wanderer will die. Should he fall in the shade or at night, his condition might well correct itself. He will no longer have to endure the external heat load from the sun. His skeletal muscles will be at rest, sharply reducing his internal heat load. In an upright human, 70 per cent of the blood lies below heart level;

in a collapsed human, all of it is horizontal and much easier for the heart to push around, especially into the head. With blood flow restored to the brain, the victim may revive. Collapsing might actually save his life, as long as he collapses in the right place or at the right time.

Andrew Hughes was looking for that elusive road to Duse, while Jane, Matt and Sam headed back to the car to await his return. Hughes pressed on through the day, absorbing a high solar heat load and sweating profusely. He wasn't yet very thirsty. Responding to the heat, his surface blood vessels were fully dilated, turning his skin bright red, even in those places protected from the sun. He felt nauseated, suffered from abdominal cramps and was growing weaker with each step as his hypothalamus diverted more and more blood away from his stomach and muscles. By nightfall, he was confused, disoriented and hopelessly lost. He collapsed in the sand.

The cool night air and his recumbent position brought him back to life. Though he was still exhausted when he awoke in the morning, his first thoughts were of his wife and children waiting for him, depending on him. He summoned enough strength to go on, and staggered on to a road just as a Tunisian farmer was driving by. The farmer took him inside his truck and gave him some water. Once his body and mind were reconstituted, Hughes, along with the farmer, set off to look for the car. They found it two and a half hours later, but Jane, Matthew and Sam were nowhere to be seen – nor was there any sign that they had ever returned there.

The farmer pulled the car out of the sand, and Hughes climbed into it, started it up and began to follow the farmer's truck. After a few miles, a tyre blew out and the car again ground to a halt. For some reason, the farmer never turned back to look for the car behind him. Hughes frantically sounded his horn to no avail, as the truck gradually disappeared. The rented car had no jack. For the second time, Hughes was alone, with no water, no idea where he was and no family. Once again, he set out on foot in the blazing afternoon sun with no clue where to find refuge or help. That night he curled up under a bush and slept. At dawn he rose to continue walking. Burning with thirst, he stumbled across an old stone well, but the water was too far down to reach and the bucket rope had rotted away. Hughes collapsed again, from dehydration, exhaustion and despair over the plight of his family.

Hughes was probably only a few hours from death when a patrol of Tunisian soldiers happened upon him. They revived him, then informed him that they had already found the rest of his family – too late to save them from the desert. Hughes' wife and sons had not been able to find the water tanks that were 'just down the road,' nor had they, as Hughes already knew, made it back to the car.

Exactly what happened to the three of them in their final hours will never be known, but we can reconstruct what happened inside their bodies as they slipped into heat exhaustion, then into heat stroke, and finally died. Subjected to relentless solar bombardment, they endured a steadily rising heat load. Whether they were seeking the shelter of their car or resting on the desert floor, they perspired profusely, for their bodies' primary task was to maintain internal temperature. All three endured enormous water loss; they had a woefully inadequate 1.5 litres and no way to replenish it. Their internal water supply dropped below the critical level; the volume of their blood decreased, becoming a salty sludge too thick and slow-moving to make it into the tiny end vessels that nourish individual cells, or to flow through the fine blood-cooling network just beneath the skin. Soon they lacked the water to sweat. Their skin became hot and dry – the telltale sign of heat stroke.

A human in the desert sun without thermoregulatory defences is no different from a piece of meat roasting in an oven. Surrounded by air temperatures that routinely reach 49°C, he will collapse, only to then be baked by sand temperatures that are 11°C to 22°C higher. The result is intense heating. In a desperate attempt to the keep the body temperature from rising that fatal 5.5°C, the hypothalamus relinquishes muscular control of surface blood flow, allowing vessels to dilate fully. The skin flushes from the increased blood that pools there, further diminishing the supply to the vital organs. The body is breaking down.

Chaos reigns as organs overheat. The stomach and intestines stop digesting, and the liver stops neutralizing their toxic by-products. The blood cannot filter through the kidneys to prevent the buildup of toxins. Muscles convulse in large, violent spasms. As the brain's delicate and intricately paced chemical reactions are speeded up by the relentlessly rising temperature, the mind and body grow confused. The hypothalamus loses control of itself and of the entire thermoregulatory system. The steadily rising internal heat alters proteins; cell membranes become distorted and porous; salt leaks in. The cells swell and burst, their contents exploding into surrounding tissues, where they cause more damage and inflammation, leading to more swelling and bursting in an accelerating and now unstoppable fatal chain reaction.

Jane, Matthew and Sam Hughes had been unable to survive what should have been a one-hour walk back to their car. Travelling over terrain with no landmarks, no water and no shelter, they were easy marks for the pitiless sun, which killed them within two days. In the same desert, five years later and under similar conditions, Mauro Prosperi survived for nine days before being rescued, drinking by his account almost no water, and covering a distance of 209 km. Can his story possibly be true?

The morning after that fierce sandstorm, Prosperi climbed to the top of a high dune. He saw no trace of a trail, nor could he see a support truck or a camp. The race manual was clear about what to do if you get lost: stay put and wait for rescue. Prosperi played by the rules. Toward the end of the day, a search helicopter passed overhead. He waved at it frantically, but it didn't see him frying in the sun. He drank his last bit of water, and later, to recycle his fluids, he urinated into the empty bottle. Then he fell asleep in the cooling night at the top of the dune.

In the stillness the next morning, Prosperi gazed out at nothing but sun and sand. There was no movement in the sky or on the horizon. Lacking water or shade, he knew the sun would most likely kill him within a day or two and decided he could no longer stay put. After wandering aimlessly for a few hours in the sun, he spotted a small structure off in the distance and made his way toward it. It turned out to be an empty Muslim shrine. The same imperative to get out of the heat of the day had inspired other species as well: he discovered he was sharing his shelter with a colony of bats.

Prosperi suddenly had a great deal in common with other desert animals. Thirst and hunger are powerful motivators, and the Italian policeman's survival instinct was activated. He climbed up under the roof, grabbed two sleeping bats and twisted their heads off. After sucking his prey dry, he ate them raw. Two days in the desert had turned a marathon runner into an opportunistic predator.

At dawn of the fourth day of Prosperi's time in the wilderness, a plane flew over the shrine. It did not spot the Italian flag he had taken from his pack and hung on a pole outside nor the SOS he had traced in the sand. By noon the sun had baked his disappointment into despair. He became suicidal, slashing his wrist with his survival knife. But his dehydrated blood was so thick it oozed and soon clotted. As the sun set, the air cooled his brain, sharpening his conscious will and with it, his instinct for survival. He could see mountains along the horizon. Remembering that the finish line was located at the foot of a mountain range, he set out to reach them.

Trying to outwit his adversary, Prosperi walked in the early morning, before the sun could muster its full power. He shielded himself against a cliff, within a cave or beneath a tree in the afternoon, then resumed his march in the evening. At night he dug a pit in the sand to keep warm. His survival thus far had been a unique combination of primal animal behaviour mixed with the props of civilization he still carried. He slaked his thirst by chewing on towelettes and by licking the morning dew off hollows in rocks. In a dried-up riverbed, or wadi, he dug up some grass and sucked the still-wet roots. He drank his own urine but saved some of it to boil a packet of freeze-dried food on his portable burner. He ate beetles and plants, and one mouse, which he killed using a slingshot made from a stick and a bungee cord. Every day he was progressing steadily east toward the mountains – except they were the wrong mountains.

On the fifth day, Prosperi spotted water dead ahead. He moved toward it hopefully but with tempered enthusiasm, aware that it might not be real. Sure enough, the water always seemed to evaporate just before he got to it, and he was never quite able to get his feet wet. It wasn't his brain that was playing tricks on him, though, it was the desert atmosphere. A mirage, unlike a hallucination, is an optical phenomenon that exists outside the brain. It can be photographed. A ray of light will bend as it crosses the boundary between two transparent mediums of different densities. Imagine a pencil as a beam of light, and picture how it looks from the side in a half-filled glass of water. This is how a lens bends and focuses light, and why fish underwater are closer than they appear to be (ask Steve Callahan). The same phenomenon occurs when a large layer of air wrapped over the desert surface (or a small pocket of air hovering over a hot asphalt road) is superheated, expanding drastically and becoming far less dense than the air layer above it. The radical difference in densities bends the incoming light rays so severely that they are nearly parallel to the ground by the time they reach the observer – that is, the desert wanderer staggering over the sand. The image projected into his forward-looking eyes actually originates from the light overhead. Brains interpret light rays as if they were travelling a straight path, so the patch of sky above is seen as a pool of water in the sand ahead.

Not until the eighth day did Prosperi stumble into a wadi that contained a real puddle. By this time his mouth and throat were so swollen that he could not swallow; he vomited his first drink. Only by taking a small sip every few minutes could he keep the water down, so he lay alongside the puddle all day and night, periodically licking at the muddy liquid. He set off again the next morning. A day later, he came across fresh goat droppings, then small human footprints, then finally the eight-year-old Tuareg girl who was making the footprints and tending the goats. The girl screamed at the sight of the desiccated carcass shambling toward her and ran off across the dune. Soon she reappeared with her grandmother, who led the poor stranger to their encampment.

Prosperi had crossed into Algeria. He was taken first by camel, then by truck, to an Algerian military hospital, where doctors reported that the desert wanderer had lost 15 kg and that 16 litres of intravenous fluid were needed to replace his water loss. His kidneys were barely functioning, his liver was damaged and he was unable to digest food. His eyes had sunk back inside their sockets and his skin was dry and wrinkled. He looked like a tortoise. But he would survive.

Military doctors in Morocco said that never before now had they seen anyone survive in the Sahara without water for more than four days. Maybe they still hadn't. Prosperi received a hero's welcome when he returned to Italy, but his tale was soon challenged by doctors, who argued that his story was physiologically impossible. They postulated that he must have been taken in, if only temporarily, by some desert nomad, then, somewhat restored, re-entered civilization with a dramatic story to tell.

Exactly how long a human being can survive without water is not known. Gathering such data would mean engaging in atrocity. The Nazis, notoriously, carried out just such experiments but never published the results of their criminal activity. Egyptian soldiers in the Sinai during the 1967 war with Israel received 3 litres of water per day, but the army still suffered numerous heat-related fatalities. Israelis, who were required to drink 10 litres a day, reported no cases of heat deterioration at all. So unquestioned is the need for large amounts of water that developing heat illness in the Israeli Army is punishable by court-martial.

Could Prosperi have survived what he said he did? A loss of greater than one-fifth of one's body water is usually fatal; he had lost one-third of his. Nine days with virtually no water far exceeds the known limits of the human body. As awesome a machine as the human body is, under extreme environmental stress, can it kick into overdrive and become even more awesome?

For humans, the desert can be thought of as a pathogen like a virus or bacteria – a natural substance that causes disease. To some extent, therefore, humans can be 'vaccinated' against the desert. Vaccines work because they expose the body to an inoculated pathogen in a weakened form, causing the body to undergo protective biochemical changes that create immunity if the pathogen should ever strike in full force. It is possible that Prosperi survived because his first four days of controlled marathon running exposed him to the desert in a weakened form and the stress served like an inoculation, one that the hotel-sheltered Hughes family never had the opportunity to acquire. Given time, the human body, ever adaptable, can be stimulated to make some defensive changes in response to any extreme environment. Fending off the heat load is the first priority in responding to the pathogen of the desert. Over the course of several days, sweating becomes easier, beginning soon after the body has begun to exercise and occurring at a lower outside temperature; this serves to keep the body's engine cool before it even starts to overheat. The volume of sweat also increases. To prevent wholesale loss of its precious minerals and salts, the body withdraws them from the sweat, which no longer stains or even smells bad, having become nearly pure water.

This increased sweating response requires water, and lots of it. To conserve that precious commodity, the hypothalamus begins periodic activation of the sweat glands within each patch of skin. Sweating becomes cyclical. Three hundred thousand sweat glands, all with their valves wide open at the same time would cause the skin's surface to flood, and water that drips off the body before it evaporates is a complete waste. To give each released droplet the time and space to evaporate, adjacent valves open and close sequentially, so that the body obtains maximum cooling from each drop of water it sacrifices.

Sweating more effectively and efficiently after several days of desert conditioning, Prosperi's skin would be cooler than that of the average wanderer suddenly cast into the desert after a week of hotel air-conditioning. He could maintain body temperature with less surface blood flow, and by now he would have more total blood flowing. Even before he became water-deprived, his hypothalamus had responded to the prolonged heat exposure by sending hormones to the kidneys, signalling them to extract even more water from the urine and recycle it to the blood. Concentrated urine and efficient sweating meant that more blood was available for Prosperi's vital organs, helping him avoid stomach cramps, muscle fatigue and fainting.

The brain is the vital organ most important to protect. An unconscious human is defenceless. The brain must be kept cool (figuratively as well as physiologically) and, conveniently, it is located close to the face, an area with a rich blood and sweat gland supply and thus well suited to dissipate heat. Normally, blood from the brain flows outward to the face. When the brain temperature gets too high, however, the direction can be reversed, and face-cooled blood will flow inward, a priority circuit that gives the brain precious, additional cooling. The evidence for preferential brain cooling in humans is controversial, but it is a well-established phenomenon in many desert mammals. Under stress and breathing hard to escape a hyena, an antelope will send the moist, air-cooled blood from its nose directly back to its brain, allowing the rest of its body to overheat while keeping its vital command centre cool. It makes sense that humans like Prosperi, functioning under stress, could invoke the same physiology.

Stress signals a danger to survival. Whether the peril is actual or just imagined, whether you're lost in the desert and short of water or trapped in traffic and late for an appointment, the body releases the same stress hormones. Adrenaline will make your heart beat faster and start you sweating. Cortisol and catecholamines will mobilize glucagon stores from the liver and muscles and convert them rapidly to glucose, the body's sugar fuel, for a quick infusion of energy. That energy burst can get you over the next sand dune or make you yell and slam the car horn.

Other reactions to stress are more specific and far more subtle. In response to too much heat, the body changes at the molecular level. All proteins and all cells have a three-dimensional folded shape that unravels at high temperature. If the heat to which they are exposed rises only gradually, however, once the temperature reaches 41°C, the cells in any organ will produce a new class of substances called chaperone, or heat-shock proteins, which bind to normal proteins to prevent them from deforming. They can even fold damaged proteins back to their original shapes. Chaperone proteins require about one hour to form and allow the body to withstand an extra 2.3°C of heat.

Heat shocking is not the only way to create these proteins, nor is heat protection their only function. Chaperone proteins are more accurately thought of as all-purpose stress proteins, because they develop after any intensely stressful 'insult' and, once formed, will defend against any other stress. Chaperone protein levels rise in response to cold, starvation, sleep deprivation, toxins and even vigorous exercise, which may be the biochemical basis for why a vigorous physical workout relieves mental stress, and why subjecting military recruits to physical hardships during basic training makes them tougher soldiers.

Chaperone proteins are produced just as easily by psychological stresses, such as isolation or fear. Consequently, there is a biochemical way in which confronting fear can sometimes strengthen the body enough to enable it to prevail over the harshest environments.

In his struggle to stay alive, Prosperi was making use of all these defences and more. A week of heat exposure had already pushed his body into full survival mode, but not many desert wanderers are also marathon runners, and that gave him some additional advantages. Desert survival depends not only upon maximizing heat loss but also upon minimizing heat production. Prosperi had a fine-tuned, highly efficient body that could produce a lot of work while generating only a little heat. Endurance training strengthens the heart so that less pumping is needed to maintain blood flow. This eases the work of breathing. Endurance training also enhances the liver's ability to convert stored energy (glycogen) to fuel (glucose). A marathon runner's muscles cover distances more easily. The total energy saved dramatically reduces heat buildup, since the heart, lungs, liver and muscles are the four biggest furnaces in the body. Endurance athletes have a preponderance of type I 'slow-twitch' muscles, which contract smoothly and steadily and thus conserve energy. Non-athletes (and more especially, athletes in sports that require fast, powerful moves such as weight lifting) have a higher percentage of type II 'fast-twitch' muscles, which burn fuel far more extravagantly.

The marathon runner's strong heart and extensive network of blood vessels make it easier for blood to flow to the surface for cooling and to reach the deep organs for nourishment. Circulation is also aided by muscular activity in the legs that pumps blood up from the veins, which explains why it is so much more uncomfortable to stand in the heat than to walk, and why a punishment for soldiers during World War II was to make them stand in the sun until they passed out. An athlete's ability to keep moving greatly favours heat dissipation by creating wind over the skin. The 'cooling down' period that runners need before they can actually stop after a race helps ensure that air flow continues until enough of their body heat has been lost by convection. Even when not moving, a marathoner has passive heat loss advantages; he has very little fat insulation and is usually slightly built – a shape that translates as a high surface area in proportion to body mass. Thinly insulated exposed surfaces will undergo rapid heat loss by conduction and radiation.

Endurance athletes have still one more card to play. For years, their steady, prolonged exercise programme has put unrelenting upward pressure on the maintenance of their body temperature. Whether the heat load is generated by muscular activity or by solar radiation, the adaptation required to sustain it remains the same. In addition to the constant buildup of heat-shock proteins, the bodies of athletes undergo a long-term change in the enzymes that control the rates of biochemical reactions. Genes are activated that alter the shape and amino acid content of metabolic enzymes, making them more resistant to heating and thus better able to maintain proper reaction rates in the face of elevated temperatures. Small molecular changes can produce dramatic effects. For example, heat-tolerant bacteria called hyperthermophiles contain many proteins similar to those found in humans, yet with only a few amino acid changes they are able to survive in boiling water. A 5.5°C rise in body temperature is usually fatal to humans, yet experienced marathon runners develop a heat tolerance that enables them to perform quite well during competitions in which their body temperatures rise by over 4.4°C. In that way, runners have become like camels. During the day, a camel's body temperature can rise 6.6°C. The animal 'stores' the heat until night-time, when it will be passively dissipated by the cool air. Avoiding the need to sweat or to increase blood flow to the skin saves water and energy, reducing heat production. A conditioned marathon runner employs the same tactic, bringing him one step closer to becoming a denizen of the desert.

To adapt to their extreme environment, however, desert

animals rely less on their physiology than on their behaviour. The heat-stressed hypothalamus in mammals, while orchestrating a defensive response from blood vessels, sweat glands and heat-shock proteins, is also signalling the cortex to go on the offensive, inducing behaviour that can dramatically reduce heat load by modifying the surrounding micro-environment – from loosening clothing to seeking or creating shelter. No matter how complex the response, when this behaviour is triggered at the subconscious level, we call it instinct. But if signals also reach the conscious mind and create the painful awareness of being 'too hot,' then the animal or human will additionally be driven by the will to survive. Life-saving behaviour for a species accustomed to modern technology might be as simple as a speedy retreat to an air-conditioned hotel, but in extreme environments, the links to civilization are easily broken. For the Hughes family, the break was as easy as getting lost on a desert road. For Mauro Prosperi, it took only one brief sandstorm to separate him from the elaborate artificial contrivance of the desert marathon and plunge him into a far more natural, elemental world, an arena in which his life depended on the resilience of his body and the activation of his latent primitive instincts.

Prosperi was put on the same playing field of survival as every other desert animal. Like them, he had to reduce heat production and increase heat loss. Any physical activity consumes energy and generates heat; that cost has to be offset by longer-term benefits in energy conservation and heat dissipation. Moving into the shade is cost-efficient – it can lower the temperature by 11°C to 17°C – but that still may not provide enough protection, especially against the solar radiation reflected up from the much hotter dry surface sand. Prosperi was lucky to find not just shade but that shrine, whose floor insulated him from the sand and whose walls blocked the wind. The shrine provided the same shelter for him as it did for the bats.

Once a desert animal, especially a big mammal, is on the

move, shade is hard to find. Camels, whose desert ancestry goes back millions of years, have developed a portable sun-blocking system called fur. Their mat of individual hairs deflects solar radiation while still allowing cooler air to circulate underneath and also leaving space for sweat to evaporate. Prosperi did not have their adaptive advantage, but he did have a human invention that was designed for precisely the same purpose: a loose-fitting Lycra running suit.

Animals without good thermal protection, such as hyenas and snakes, deal with the heat problem by being nocturnal. Without low-light vision or infrared sensors, night travel would be dangerous for a human. Prosperi compromised. He walked in the early morning and late evening, resting during the heat of the day behind a bush, a rock, a dune – anything that would cast a shadow.

The lack of moisture in the air and sand means that the desert is unable to retain heat once the sun disappears. Ground temperature can vary over 50°C from day to night. One of the best energy investments a desert inhabitant or a human visitor can make is to dig a hole in the sand. At a depth of 2 metres, sand provides enough insulation to maintain a nearly constant temperature of 10°C to 16°C. Aardvarks, jerboas and toads burrow into the sand to sleep; and so did Prosperi.

The extreme temperature fluctuations that forced the runner and the other animals underground also generate a daily supply of water, if you know how to collect it. Air, even desert air, always contains some moisture. Since cold air cannot hold as much moisture as hot air, the rapid, drastic drop in temperature at night causes what little water there is in the air to precipitate out. The water condenses on rocks and surface sand before being evaporated by the morning sun. Desert beetles know how to collect it. They face the cool morning breeze and raise their concave backs; water that condenses there trickles down into their mouths. Prosperi collected some of the condensed water by licking the morning dew from crevices and depressions in rocks. He would have done much better had he carried a large plastic sheet with a hole in the middle and placed it over a pit in the sand. Putting rocks around the edge of the plastic to prevent it from falling in, and adding one rock toward the centre to weigh it down, he would have created a funnel to channel condensed water down the hole and into his water bottle positioned underneath. The desert is stingy; the overnight collection would not have given him enough to live on, but he could have added the drops to whatever else he found each day.

Prosperi didn't have that technological advantage, so he was reduced to trying to get water the same way desert animals do. Besides licking rocks, he pulled up any rare tuft of grass he came across and sucked on the roots. He followed wadis and, the day before his rescue, found one that still contained a small, muddy puddle. He drank like a camel and, like a camel, lay by his watering hole overnight.

What little water Prosperi found during his ordeal he recycled by drinking his own urine. This might seem like taking poison – and in some ways it is – but it helped prolong his survival. Salts in the urine never become as concentrated as they do in the blood. Putting the liquid back in the blood therefore results in relatively more water being added than salt, a net increase in fluid volume. Because the urine will contain a high concentration of urea – a toxin that will eventually damage the kidneys – it can only be drunk a little at a time. So while it might be okay to drink urine, it can very quickly become too much of a good thing.

That caution applies to food as well. Like any other carnivore, Prosperi tried to hunt, but he was lucky not to have been too successful, for while he may have had the same hunger drive as the animals around him, he didn't share their metabolism. Animals as predators possess heat-resistant enzymes that digest other animals more efficiently and excrete the byproducts with less water loss than humans do. Animals as prey have high fat and protein content; eating them requires more energy and gives off more heat than eating plants, which are mostly ready-to-digest carbohydrates. That's the reason why human visitors to a hot climate eat less and often feel an instinctive aversion to meat and fats.

Humans can go a long time without food, and if they are without water, they should. Even digesting carbohydrates consumes water, and all foods contain at least some salt, which dries out the body even more. Travellers lost in the desert don't die of hunger; they die of thirst as they wander around with no chance of finding their way out.

Mauro Prosperi was hopelessly lost in a wilderness with too much vacant space and too few landmarks. Yet navigation is not a problem for seabirds that can cross entire oceans following lines of magnetic force. Adding insult to injury, it's not even a problem for desert ants; their photoreceptors enable them to follow patterns of polarized light in the sky. Hapless humans, however, are oblivious to magnetism and polarization, and so, if we have neither a compass nor knowledge of stars, we're clueless when it comes to large-scale navigation. Prosperi's only hope for survival in the Sahara was to stumble across someone capable of rescuing him, and he did – that eight-year-old girl. The marathoning Italian policeman and the goat-herding Tuareg child stood on the same barren patch of sand at the same moment. Both were human beings, sharing roughly the same physiognomy. Yet the little girl was thriving, while the marathon man was dying. The difference in their well-being at that moment was a result less of adaptations in their bodies than in their brains. Over the centuries, Tuaregs have learned how to traverse the desert by carrying simple tools, following basic landmarks and knowing where to find water and shelter. At first, Prosperi's desert crossing was no different. He carried a backpack, followed a marked trail and knew that there would be water and tents at every checkpoint. But he was guided by the rules contained in his race manual. Tuaregs are guided by the teachings of their elders. Prosperi's existence was dependent on the Marathon des Sables – a thin, artificial culture temporarily laid over the desert. One brief sandstorm blew away that net of support. Once exposed to the elements, he was barely able to survive nine days. Tuaregs have thrived in the same environment for generations because their deeply rooted culture arises naturally from their surroundings, making it much less vulnerable to the vicissitudes of the desert. Pushed to the limits of survival, both will resort to inbred animal instincts. Tuaregs, however, have many more layers of protection before being forced to that level, and they all originate in the mind.

Humans in the desert are physically challenged compared with their animal neighbours: they're too sensitive to heat, they can't store enough water, have limited endurance and can't tune in to natural navigation systems. They do possess, however, a supreme capacity to learn by observation, to store and transmit information using language and to integrate and apply their knowledge by turning survival skills into customs. In other words, they overcome their handicaps with their brains.

Desert people have learned to mimic the anatomy of the other large but far more adept (and adapted) desert mammal, the camel. Lacking adequate fur, nomads cover themselves completely in loose-fitting robes, jalabas, which deflect sunlight, allow air to circulate and leave space for evaporation. Born without the camel's padded feet to insulate them from the hot desert sand, nomads cover their bare feet with knitted wool socks. Camels have dense hairs in their nostrils and ear canals, and a double set of long eyelashes to keep windblown sand out of their noses, ears, and eyes. Humans make up for their meagre hair supply by wrapping their heads with shawls, leaving exposed only a narrow slit for their eyes. A camel's nose has a large hollow chamber in which air is cooled before it is exhaled, allowing the water that condenses out to be recycled. Tuaregs sleep with scarves across their mouths to trap the moisture in exhaled air so that they can breathe it in again.

Even with all our adaptive ingenuity, a human's performance in the desert falls far short of that of a camel, some of whose physical features simply cannot be replicated. A nomad can still profit from them, however, by domesticating the animal and exploiting its capabilities for his own ends. Camels are unequalled in their ability to survive passages across vast stretches of desert without water or shelter. Distant oases are reachable for nomads only because they have camels to bring them there. Camels store huge quantities of water. In a few hours they can drink 227 litres, which they distribute not – as children are sometimes told – in their humps or any other specific place but all over their bodies. This water is not accessible to humans, but it does keep the camel going a long time, turning it into a mobile carrier on which can be hung gourds filled with water that is quite accessible.

Camels are also portable shelters. They are large but narrow-chested animals that, if given the choice, will orient themselves parallel to the sun's rays to minimize radiation absorption. If the beasts are led perpendicular to the sun's rays, however, nomads can walk continuously in their broad shadows. The bulk of a camel's body is supported several metres off the ground on long thin legs – an evolutionary trait that minimizes contact with the layer of air immediately above the sand, which is superheated during the day and supercooled at night. Temperatures 2 or 3 metres off the ground are far more moderate. With a camel to ride atop during the day and to use as a sleeping platform at night, any nomad will be ensured a more pleasant journey.

Still, the desert traveller and even the desert dweller live in constant fear of dying of thirst or of getting lost. They must fully exploit every advantage their brains can give them to minimize the risk. They must make utmost use of their power to plan ahead and their capacity to remember the distant past. On journeys when they have water to spare, nomads might collect ostrich eggshells, fill them with water and bury them along the route, knowing that the next time they pass this way they could be desperately thirsty and that the prepositioned caches might save their lives. The water won't be useful, however, if it can't be found. Nomads can't detect magnetic fields or sense polarized light like birds or ants, but they can see paths through the desert using maps that exist only in their minds. They can recognize subtle landmarks they have never seen before that will guide them to places they have never been. They have internalized the instructions passed down to them by previous generations, and so are able to follow the invisible tracks left by their ancestors.

These adaptations are uniquely human, although nomads share many other behaviour patterns with their desert cohabitants. Some of the duplication is the result of learning to imitate animals' survival methods, but much of it is generated by a common instinct that transcends individual species. Responding to the same environmental threat, animals and humans are pushed to develop the same strategies to defeat it, but humans often go one better. All desert creatures look for water along dried-up wadis. Kalahari bushmen, however, know that even when wadi grass roots are dry, they should be buried in a hole in the warm sand to generate a little moisture. It can then be sucked up through a hollow reed inserted into the hole. There may also be some water to extract from grassless wadis. Current slows at every bend in a river, especially around its outside curve. Some of the water that collects at these bends will percolate down into the sand and may remain there long after the surface has dried up. Digging up the subsurface sand along an outer bend, then placing it in a cloth and squeezing it, may yield a few precious drops of water.

Chasing food is often a losing proposition in the desert. The pursuit generates heat, and the prize may not provide as much energy as was lost catching it. Most desert predators limit their chases to short sprints. Nomads never run at all. They catch some prey by setting traps, but mainly keep their meat supply close at hand by travelling with goats and sheep. The prized desert dessert is a date. It's the perfect hot weather food, 70 per cent carbohydrates and the rest water. Animals and humans alike are attracted to them. Animals lucky enough to come upon a date-palm oasis will eat their fill, and so will humans, who will then pack more to take along on their journey. And before leaving, they will replant some of the seeds to provide more 'luck' for next time.

Jerboas and Tuaregs in the Sahara, and desert toads and opal miners in the Australian outback, all go underground to escape the heat. The jerboa, a jumping rodent with long hind legs and a long tale, estivates – a kind of reverse hibernation in which it spends the entire summer safely beneath the surface of the sand. Tuaregs build houses with basements 9 metres deep where they can take refuge when the heat is overwhelming. Australian toads escape the blazing heat of the outback by burrowing well below the surface. They store water in their bladders and can live underground for years. Opal miners also go underground, using digging machines to carve out enormous caves for themselves and their families. These caves have their own water supply and sometimes even swimming pools. Though the above ground temperatures can exceed 49°C, their house temperature remains constant at about 24°C. The town the miners founded, Coober Pedy, was given its name by the local aborigines who refer to the place as *kupa piti*, which means 'white man in a hole.'

Humans survive mostly by using their nimble brains to adapt the desert to their bodies, yet there are indications that they can also adapt their bodies to the desert, albeit at a much slower pace. Heat sculpts body parts. An obvious example, one that developed in mammals long before there were humans, is the positioning of testicles. Since the sperm cells in testicles cannot tolerate the internal temperature maintained by warm-blooded animals, a thin pouch evolved to hang them outside the body. Suspended in the scrotum, testicles can be cooled by air flow and, at least in four-legged animals, blocked from the sun's rays. Humans have actually interfered with this adaptation by standing up and putting on clothes. Clothes retain radiant heat and block convective heat loss, increasing the temperature inside the scrotum. 'Civilized' males have disordered cell patterns in their testicles compared to those in males who go naked.

Heat also sculpts body shapes. The geometric shape that has the smallest surface area in proportion to its interior mass is the sphere; if internal heat can be dissipated only at its surface, spheres will retain heat the longest. The shape works well for Eskimos, who tend to have well-rounded bodies. Desert dwellers are better served by a tubular or angular design, providing far more surface area for cooling. People of the African plains, like the Masai and Samburu, who evolved in hot dry climates, are tall with long arms and legs. To favour heat loss further, they have only a thin layer of insulating fat under their skin.

Tall and thin may be good for heat loss, but if there is no place to store fat, it is a dangerous shape in a barren wilderness, whether a desert of sand or of water. The Polynesians' ability to store fat allowed them to survive long, uncertain sea voyages, and desert nomads could use some insurance too. If their body fat were concentrated in one place rather than spread out evenly, it would be readily available for energy production yet still have a minimal insulating effect. Camels have perfected this idea by developing not just long slender legs for heat dissipation but humps in which to store their body fat. Similarly, the long-limbed Hottentots of South Africa are able to store large amounts of fat in their buttocks. A proclivity for depositing fat unevenly remains widespread among humans. In deference to the demands of thermal regulation, tubular areas of high heat exchange – the head, neck, arms and lower legs – remain mostly free of extra fat. The bulk of it is targeted to the buttocks, thighs or abdomen, much to the consternation of weight-conscious individuals everywhere.

Besides relocating body parts and remodelling body shapes, heat has a say in selecting body colour. Dark skin has always been advantageous for humans in central Africa because it contains more melanin, a pigment that fends off the intense ultraviolet radiation of the equatorial sun. It gives its wearer protection against sunburn and skin cancer. But darker pigments also retain heat. In tropical areas with plenty of water, the added heat load is neutralized by increased sweating, but for humans who have migrated into arid areas, the advantage of ultraviolet protection is offset by the disadvantage of needing more water to maintain optimum body temperature. Light skin reflects more heat and lowers the water requirement but allows more radiation into the body. Evolution has apparently not yet solved this dilemma; deserts contain people of every skin colour. The ideal skin pigment would be one that blocks ultraviolet light yet still allows heat to escape. However, now that humans have invented clothing, skin pigmentation will become irrelevant to evolution. It is easier and quicker to change clothes than to change skin, and just as effective. Natural selection as a driving force of adaptation will be disconnected, and evolutionary improvements will stop.

Human beings are a work in progress, still incompletely adapted to the various environments into which they have ventured but often able to survive there through a combination of animal instincts and intelligence – their own and their culture's – that trumps their physical shortcomings. Still, how much environmental adversity can an out-of-place, disconnected human overcome? Could instinct and intelligence really have kept Mauro Prosperi alive in the desert for nine days? Prosperi's story is not unique. A century ago, a prospector named Pablo Valencia survived eight days in the Mojave Desert with one day's supply of water. He didn't have civilization, desert wisdom or even a marathoner's body, but, like Prosperi, he had animal instinct, human intelligence and a fierce will to live.

The Mojave is part of a forbidding expanse of desert that runs like a corridor through the southwest United States and into Mexico. For settlers migrating westward, it was the desperately hard final leg of the trail that they encountered as they descended the Rockies. The finish line was the Sierra Nevada mountain range of California, which blocks rainfall from the Pacific and guards the desert. The trail offers no relief until it reaches the far side of the mountains.

One particularly desolate stretch of the trail lay along the southern Arizona territory just a few miles north of the Mexican border, barren except for occasional cactus and chaparral, with an average temperature over 35°C. The only water to be found consisted of chance remnants of rare rainfall collected in potholes or rock crevices below granite ledges. The route was known as El Camino del Diablo, 'The Route of the Devil,' marked nearly every kilometre by the grave sites of travellers unable to complete their one-time, one-way journey west. Even today the region remains a death trap for migrant workers trying surreptitiously to cross the border from Mexico.

In 1905, in addition to settlers heading for California, there were isolated explorers in the region – among them a naturalist looking for the secrets of desert life and prospectors looking for gold. W.J. McGee, the director of the Saint Louis Museum, and his Papagos Indian guide, José, were camped by a water pool when in rode Pablo Valencia, a Mexican prospector, accompanied by his vaquero guide, Jesús Ríos. Valencia was returning to a 'lost mine' he said he had rediscovered, and had hired Ríos to help him stake a claim. McGee described Ríos as sixty-five years old and 'claiming familiarity with the country but erratic and inconsequent, and little dependable in any way.' Valencia was 'about forty, of remarkably fine and vigorous physique, openly scorning hunger and thirst and boasting ability to withstand far beyond ordinary men the habitual inconveniences of the range.' McGee judged him 'lightly burdened with acute sensibility, imagination or other mentality; indeed, an ideal man to endure stressful experience.'

After a day's rest Valencia and Ríos set out on horseback before dawn. Sometime after midnight, Ríos came back alone but with both horses, reporting that after riding 56 km, Valencia had 'sent him back to re-water the horses' while Valencia kept a 9-litre canteen of water and continued ahead on foot into the trackless desert. They had agreed to rendezvous the next day on the far side of some mountain. McGee called it 'an inane if not insane' plan.

Ríos left the next morning but returned again that evening, declaring that he had been unable to find Valencia. McGee sent out José, an expert animal tracker, to look for Valencia, directing him 'to go to the limit of his horse's endurance and then to his own limit beyond.' José easily found the trail and followed it through the day and into the night. When his horse became too tired, he continued on foot until he could go no farther, returning to his horse at dawn and then back to camp where he collapsed in exhaustion. McGee and José agreed that any further attempts at rescue, besides being dangerous, would be pointless. Valencia had now been out in the desert over three days with but one day's supply of water. The naturalist and the Indian were both sure that Valencia was already dead.

Ríos rode away, and McGee and José uneasily resumed their normal camp routine. Eight days after Valencia's disappearance they were awakened just before dawn by a strange noise that at first they took to be the bellowing of a lost bull steer. They ran out to investigate and, at the base of a dry gulch, came upon 'the wreck of Pablo.' McGee described him as 'stark naked; his formerly full-muscled legs and arms were shrunken and scrawny; his ribs ridged out like those of a starving horse; his abdomen was drawn in almost against his vertebral column . . . His eyes were set in a winkless stare . . . able to distinguish nothing save light and dark.' His ears were 'deaf to all but loud sounds . . . His lips had disappeared' so that 'his teeth and gums projected like those of a skinned animal, but his flesh was black and dry as jerky . . . Even the freshest cuts were as scratches in dry leather without a trace of blood.' In short, McGee provided a very acute description quite similar to the one provided by the military doctors who treated Mauro Prosperi. Both McGee and Prosperi's doctors estimated that their patients had lost 25 per cent of their body weight.

Valencia needed water desperately, but he was unable to

swallow, so McGee rubbed water all over his body. At first it rolled off his skin as it would off tanned leather; then the skin began 'absorbing it greedily as a dry sponge.' Soon he was able to drink, and in a few hours his circulation was restored enough for him to bleed from his numerous cuts. Blood-borne immune cells were now able to reach his open wounds to fight infections that already had several days' head start, and McGee observed a rapid and fearful swelling and inflammation of Valencia's limbs. With the essential ingredient of water, plus some food, all Valencia's body systems were gradually restored, though McGee reported that the Mexican prospector's once black hair remained permanently grey. In three days he was well enough to recount his ordeal.

After separating from Ríos, Valencia had continued on foot, slept in the sand and reached his mining site the next morning. He set about collecting samples and posting notices for a mineral claim, finishing his work before noon. He drank the last of his water shortly after, expecting to rendezvous soon with Ríos. When he failed to show, Valencia headed out for a road Ríos had described to him. By nightfall he had found neither Ríos, nor the road, nor any water. He soothed his dry throat by gargling his urine. The following day he found some trails that led nowhere and an immense pothole, completely dry. Trying to relieve the heat and reduce his weight, Valencia threw away his clothes and even his shoes, keeping only the canteen so that he could save his drops of urine, which he was now drinking.

By the fourth day Valencia became convinced that the trail he was searching for didn't exist, that Ríos had never had any intention of meeting him and that Ríos' plan all along had been to let the desert kill Valencia and then return to stake the mining claim himself. Valencia rued throwing away his clothes, not so much for the loss of sun protection but because in his pocket was the knife he would have used to stab Jesús Ríos. His desire for vengeance became an obsession, giving him the will to continue. He walked in the mornings, rested in the shade of a bush to avoid the afternoon heat, then walked again in the evenings. At sunset he caught flies and spiders and one green scorpion, which, after he ground off its stinger with a stone, yielded a few precious drops of moisture. At last Valencia came across the El Camino del Diablo wagon trail. There were no settlers travelling on it, but he followed the ruts, knowing they would lead back to McGee's camp. Several times he collapsed on the trail but was inspired to go on by thoughts of Ríos and the pleasure he would take in knifing him.

Valencia's recollection of his final day on the trail was vague except for one experience that he described clearly: as the sun rose that last morning, he died. The sun burned down on his body, but his soul was reluctant to abandon it and hovered over him the whole day. When darkness fell and the desert cooled, he felt a shadow stir his naked body and push it along. Whatever it was lay beyond his control. Just before dawn he was surprised to hear a bellowing noise coming from within his body. He crawled a little farther on and was met by José and McGee, who had heard the noise.

Pablo had lived eight days in the desert, seven of them without water other than the few drops he extracted from the scorpion and his recycled urine, which dried up after five days. His survival would seem to be a physiological impossibility, yet, as McGee was to write, 'his wrath spurred him on – a potent incentive which carried him miles and doubtless saved his life.'

Valencia was convinced he died on the last day though his body kept going, not even losing its trail sense. Did his out-of-body experience represent a disconnection of his cerebral cortex, taking away his will? At the moment his endurance reached its limit, did his higher functions relinquish control, diverting it to the lower brain in order to save power? It seems Valencia was able to energize his most basic animal instincts for a last desperate, and successful, attempt to save himself.

Pablo Valencia was conditioned to the desert. He was a hardened prospector, physically fit, and powered by vengeance. Mauro Prosperi was a world-class endurance runner, acclimated to the heat, inured to physical hardships through years of training and discipline. Each man was able to deploy a fine-tuned body defence and activate a fierce will to survive, adding human intelligence to latent animal instincts.

Pablo didn't die and Mauro didn't die – and Mauro didn't lie. His body corroborates his story. The condition in which Mauro Prosperi was found, and in which Pablo Valencia was found in a different era and in a different desert, is not explainable any other way. Prosperi's body provides compelling testimony to the kind of damage the desert can inflict and, at the same time, evidence of what the body can sustain when pushed to its extremes. He competed against the desert as a decided underdog, but he won, turning in the performance of his life.

UNDERWATER

THE PULL OF THE DEEP

MY AIR BUBBLES WERE ACTING STRANGELY. As I exhaled into my regulator they sank through the water, floated below my feet and bumped against the tunnel's slanted floor. I watched as they rolled on downward and out of sight – then realized that I was upside down.

There was no other way to tell at the moment. I was inside a flooded lava tube, a channel that once vented molten rock from an underwater volcano into the sea. The entrance was a nearly round hole 2 metres in diameter in a smooth mound of rock 9 metres below the water's surface. The tube was lined with a thick layer of fine sediment and ran as straight and smooth as a drinking straw, angling steadily downward to who knows where.

I had come upon the vent unexpectedly and entered it without a dive buddy or any standard caving equipment, such as a reel of rope to be paid out along the route from the entrance so that I could always find my way back. But the tunnel had no side branches, and I could still see the light at the entrance. Swimming diagonally downward, I had penetrated about 15 metres in, to a depth of 27 metres. The light was growing dim and I was about to turn around when I saw a sea lion coming rapidly up out of the depths. Sea lions are large, gentle creatures that often amuse me on dives in the Galapagos Islands, but this one was heading straight toward me in a narrow tunnel and collision, or at least a playful bump, seemed imminent. I held my mask with one hand and my regulator with the other, afraid to have either one or both knocked away by the impact. The sea lion loomed close enough for me to count the whiskers on his face, then deftly changed course, wiggling past me with no contact whatsoever. My manoeuvres hadn't been nearly as graceful. My dive to the floor of the tunnel had stirred up a layer of sediment that formed into a thick cloud of fine black powder. It spread widely and then hung suspended in the water, obscuring my view of the tunnel and diffusing the light enough so that I could no longer tell from which direction the light was coming. It would take hours, maybe even days, for the tiny silt particles to settle. The air in my tank would last me at least another forty minutes so long as I didn't get nervous and start breathing rapidly. There was plenty of time for me to get out of the tunnel – if I knew which way to go. I couldn't tell up from down, in from out. My weight was exactly counterbalanced by the buoyancy of the water, so I felt no downward pull from gravity. There was nothing to hear except my breathing, and nothing to see except the barely discernible rock walls of a tunnel that no longer had a top and bottom.

I rotated myself until the line of exhaled air bubbles bisected my forehead, meaning that I was upright, and then followed the line, since it was pointing to the ceiling. The trapped air was pooling in a crevice of the rocky wall. Some seemed to be escaping off to one side, however, which meant that side of the wall was higher, and since the tunnel was more or less uniformly slanted, I started swimming that way, kicking up more silt as I went because I couldn't see where I was going. I expected to come through the cloud at any moment but it extended much farther than I thought. Maybe the silt was sinking down through the tunnel and I was swimming down along with it. Just as I was beginning to think I should turn around, my ears popped. Air in a diver's ear maintains itself at the same pressure as the surrounding water. Near the surface, the ear senses less pressure and responds by snapping open its relief valve. The pop meant that I was going up. I swam ahead a few more strokes and emerged from the cloud of silt. A beacon of light, the cave entrance, led to the open sea.

What had been a leisurely dive turned into a mini-survival test because of a close encounter with a playful sea lion. He was in his natural element. I was in a hostile environment, sustained by an elaborate set of artifices that gave me such a false sense of security that I had entered without a buddy or a rope. My air tank, regulator, buoyancy-compensation vest, wet suit, weight belt, mask and fins made me comfortable underwater, but they didn't make me into a marine mammal. Closer to the limits of my survival than I had realized, I was nearly overcome by a problem that would have been trivial for any true sea creature.

Being under the water and inside the earth at the same time brings on a sense of total isolation from the outside world – a mixture of anxiety and tranquillity that draws people into underwater caves, explained my friend and fellow diver Henri Cosquer, as he began to tell me an incredible story. For years Henri has owned and operated a diving school in Cassis, a picturesque fishing village on the southern coast of France. Surrounded by white limestone cliffs that plunge precipitously into the Mediterranean Sea, creating a series of narrow fiords called callanques, Cassis became a resort after being overrun by the French Riviera.

Though I had spent much time exploring the callanques – climbing to the tops and diving to the bottoms – I never gave much thought to what might lie inside them. Neither had most other people; fewer still had the skills to find out. In the course of one of his thousands of dives, Henri came across a 2-by-1 metre hole in the rock at a depth of 37 metres. He passed it dozens of times on other dives before curiosity got the better of him and he spontaneously decided to go in. The opening was coral-encrusted, but after about 5 metres the coral faded away as the light dimmed. Henri swam on for another 18 metres, until the light from the cave entrance was a pinpoint behind him, and then turned around. It wasn't safe to go any farther – this time. But the cave was intriguing. He'd be back.

With lanterns and backup lights, Henri returned to make a series of dives. Each time he penetrated farther. He noted that the tunnel was rising steadily. At a bifurcation, he followed one branch to a dead end, 46 metres from the entrance. The next dive he followed the other branch, and eventually came to a narrow, body-sized constriction in the rock walls. Shining his light through it, he could see that it opened up on the other side. He was 152 metres inside the tunnel but only 12 metres below sea level.

Henri returned with more equipment. He was able to pass through the tight space only by taking his tank off and passing it in ahead of him, while keeping his regulator tightly clamped in his mouth. On the other side, the tunnel continued straight on another 9 metres, then the walls suddenly sloped up and out to form a vast underwater chamber. The roof was a mirror, reflecting the beam of Henri's searchlight back down at him. He knew that meant surface water and that there was air above. He swam upward, broke the surface and found himself inside a huge cave. Shining his light around him he saw that he was in a pool 30 metres wide surrounded by a rocky shore with fluted spires of limestone – stalagmites from below and stalactites hanging from the roof. He felt as if he had entered a temple of the sea. Not quite believing that any of this could be real, he swam to the shore, took a few steps on the rocks and touched the stalagmites. He said they were warm, like they were still alive. Then he dove into the pool and swam back to the rest of the world.

On the next dive to his secret cave, Henri tried to explore a little more, but both his searchlight and backup light malfunctioned and he barely made his way back through the pitch-dark cave and tunnel. The episode scared him so badly that he didn't return to the cave for three years. Perhaps he really had violated the sanctity of a sacred temple. Finally, he returned with other divers, laid in a safety rope and explored the cave further. There were no other entrances. The cave was deep inside the callanque yet above sea level, the oxygen presumably coming through twisted crevices that allowed in air but not light. In the total blackness, Henri advanced with his lantern. He stopped for a short rest and placed the light alongside him. It illuminated a small section of the wall with a strange shape on it. Henri brought the lamp closer. It was a painting of a human hand.

Henri's first reaction was that it was impossible. How could someone else have gotten to this spot with paint and brush and then carefully drawn a hand on the wall? Henri and the other divers began a systematic search, beaming up their lights to scan the walls in every corner of the cave. What they found left them stunned. There were dozens of paintings – of penguins, caribou, birds, buffalo and horses – all carefully designed in a very simple, very primitive style. Henri had indeed not been the first person to enter this place. Far from it. He was standing in a temple where humans had come to worship twenty-seven thousand years ago.

But how could they have gotten there? A layer of rock that hadn't budged in a million years sealed the roof and walls. The only access was through the floor, but to enter that way meant diving 37 metres below the surface and then swimming 164 metres in total darkness. Henri had needed full scuba gear, and even then it took him several attempts. Henri is short, stocky and full-bearded, and proud to say that he looks like a caveman – his dive boat is named *Cro-Magnon* – but he admits that even his ancient relatives would have been unable to make the dive on a single breath.

Nevertheless, breath-hold diving, also called free diving, has a tradition going back into prehistory; there is evidence that Neanderthals practised it 40,000 years ago. For at least 6,000 years it has been done professionally by women in Japan and the South Pacific, who dive holding large stones to help them sink quickly to depths as great as 30 metres. The diver then spends a minute or so collecting seaweed, shellfish, pearls and anything else of commercial value before she is pulled back to the surface by a rope tied around her waist. Ancient Greek frogmen practised breath-holding so that they could stay underwater long enough to cut the anchor lines of enemy ships. Only in the past few hundred years has man developed the technology to stay underwater longer than one breath.

The earliest efforts were diving bells – inverted wooden jars that were lowered beneath the surface with air trapped inside. Since the air was quickly exhausted, more advanced models were made with hoses connected to air pumps at the surface. A comparable system was developed thousands of years earlier by water spiders, which weave airtight silk umbrellas attached to underwater plants. They fill these umbrellas with air by trapping it between their legs at the surface and carrying it down in the form of bubbles. Underneath their air-filled domes, they lie in wait for prey.

To increase mobility for a human diver, the diving bell was reduced to the size of a helmet, and with that the era of hard-hat diving began. For any one diver, however, the era would end abruptly if he lost his balance, since his tilted helmet would immediately fill with water. Adding a cumbersome watertight suit to seal out the water made movement tortuously slow, but at least the enclosed diver could bend over without drowning. The big advance in diving came only about sixty years ago, when a demand regulator – a gas valve that responds to pressure changes – was combined with a tank of compressed air to create a self-contained underwater breathing apparatus, or scuba. Scuba diving provided the reliability and freedom of movement that at last allowed humans to travel through the most alien and least explored of all the extreme environments on earth.

The realm beneath the waves is a cold, dark, airless world, seemingly without gravity. Humans fend off these dangers with a rubber wet suit, waterproof lights, compressed air and a buoyancy-compensation vest, but as we descend farther and farther into the sea we become increasingly vulnerable to the sheer weight of the water above us. Water is more than a thousand times heavier than air. As a diver descends, he must support the weight of the water, steadily increasing by the ton. The enormous pressure pushes in on the body, compressing it evenly on all sides. The only reason the diver is not crushed is that, like the sea, his body is made mostly of water – a liquid nearly impossible to compress. But bodies have four hollow spaces that are filled with air – a gas that compresses very easily. Inhaled air is channelled in four directions. Most of it is drawn through the trachea into the lungs. Some passes to the back of the throat, where it is swallowed and works its way down the digestive tract. A small amount of air flows into the skull, via small ducts that lead to the sinuses or through larger tubes that carry it into the middle ear. If the air pressure in each space is not continually pumped up to match the increasing water pressure as the diver descends, his body cavities will very quickly collapse.

The lungs would be the first to feel the squeeze. At a depth of 30 cm there would already be nearly 91 kg of pressure on the chest wall. Another half a metre down would be the limit for inhaling atmospheric air; below that level the pulmonary muscles are not strong enough to expand against water pressure. A fleeing prisoner who hides underwater and breathes with a hollow reed in his mouth had better be in muddy water because he can only submerge a metre if he expects to keep breathing.

The supple walls of the digestive tract – the oesophagus, stomach and intestines – are also no match for water pressure and quickly collapse. That collapse does not pose a problem, however, because it is not painful, and, unlike breathing, eating and defecating, are not normally activities a diver does underwater.

Sinuses, being entirely encased in skull bones, are somewhat resistant to pressure changes, but ears, only partially enclosed by the skull, are much more vulnerable. The middle ear is a chamber containing three tiny bones connected end to end to transmit sound waves from the eardrum to nerve sensors that stimulate the brain. Air gets into the chamber through the Eustachian tube, a long, delicate channel that brings air up from the throat. With a descent of only 1 metre, outside pressure makes the eardrum bulge inward and causes pain unless more air enters the chamber and provides counterpressure. Adding another 30 cm or so of depth will contract the air space enough to make the soft lining of the Eustachian tube flutter closed. When the tube is blocked, the chamber is sealed. Should the diver ignore the pain and continue downward, the pressure would be relieved within another 3 to 4 metres and the pain would disappear – replaced by a loss of hearing and severe vertigo, as cold seawater floods into the ear through the ruptured eardrum. Humans are able to prevent the dam from bursting by utilizing the Valsalva manoeuvre, a technique to clear the ears that's very similar to blowing your nose. A diver with ear pain can increase air pressure in his mouth and throat by blowing against his closed lips while pinching his nostrils – face masks are made with soft rubber covering the nose just so that noses can be pinched from the outside without breaking the watertight seal. The air backs up into the Eustachian tube, forcing it open and inflating the middle ear sufficiently to balance the outside pressure. The manoeuvre is critical at the start of the dive, where pressure changes are most abrupt.

Hovering a few metres below the surface off the rocky shore of an island in the Galapagos, I was pinching my nostrils, waiting for my ears to clear and wondering why none of the seals zooming down past me had to squeeze its nose. Not that a seal could anyway. Not since their hands and feet evolved into flippers. Still, seals have ears and lungs, so how do they dive so much more easily than I do? Their trick is to exhale before going into the water and then let their lungs collapse under the increasing pressure. Residual air is forced into the upper airways, and that seems to be enough to maintain ear pressure. As for the collapsed lungs, the seals simply don't breathe – not a trick that would work for us. It works for them because they can easily hold their breath for over an hour, an adaptation they needed to develop in order to return to the sea, once their ancestors began breathing air. This extraordinary ability evolved through relatively minor changes in their mammalian systems.

Seals have proportionately the same size lungs as humans and use the same oxygen-carrying protein, haemoglobin – except that seals carry a lot more haemoglobin than we do. In addition to having a relatively large volume of blood, seals are able to rapidly inject extra red blood cells into their circulation right before a dive by contracting the spleen, the organ of blood storage. Some other mammals have contractile spleens; initially tantalizing evidence that the ability evolved among South Sea pearl divers, however, has now been largely disproven. Humans and seals do both have myoglobin, a protein stored in muscle that provides an emergency source of oxygen, but again, seals have a lot more of it and use it as an oxygen reservoir. An abundance of haemoglobin and myoglobin lets seals store large quantities of oxygen outside their lungs, giving them a steady supply even when their lungs are shut down. And they use this supply sparingly, because seals have a 'mammalian diving reflex' that lowers their heart rate and shunts blood to their brains as soon as they enter the water. Human adults have lost the reflex, but it is still retained in children. It explains how seals can remain motionless on undersea ice shelves waiting patiently for the right fish to pass by, and partly explains how children can 'come back to life' after being submerged for over an hour.

Seals and humans both need to maintain the same internal body temperature, yet seals can dive even in Antarctic waters, and enjoy it. They solved the heat loss problem by developing a layer of blubber around their bodies. Fat has very little blood flow and therefore provides highly effective insulation. Paradoxically, children tolerate immersion in frigid water far better than adults because cold penetrates them much faster. Their bodies have proportionately more surface area and less distance to their centres. They become 'quick-frozen' into a state of drastically lowered metabolism, minimizing their need for oxygen. Activation of the dive reflex followed by quick-freezing explains why all the cold-water survival records for humans are held by children. Adults don't have either of these advantages but, like seals, many do have thick layers of blubber around them. Though they may have less energy than fit people, they are able to survive immersion longer. So, if you're going to fall into the sea, try to do it in the tropics, preferably with your clothes on, to retain heat. Get out of the water as fast as possible, and if you're not a little kid, at least be fat.

Humans are mammals with the same basic organ systems as seals, but without all the fine-tuning that seals have undergone, the final difference in underwater adaptation is dramatic. Nevertheless, free divers do the best they can with what they've got. With rigorous training, some can hold their breath for over three minutes and make round-trips to depths of over 76 metres. The most demanding part of the training involves mental discipline and body control; it amounts to underwater yoga. The pioneer practitioner of the hybrid sport was Jacques Mayol, who became a French national hero when he established the free deep-diving record at age sixty-four. Jacques and I often swam together. As I splashed vigorously along, trying to keep up with him, he would glide ahead to the shore and smoothly transform himself into a land animal by continuing his exercise sitting with his legs crossed, placing his hands on his knees and closing his eyes. He was 'rejoining the water', he used to tell me. Jacques said he felt his ancestral aquatic origins within him and thought that, like the dolphins, humans could return to the seas. He often said he was more suited to life in the water than to life on land, and I believed him. His sensitive soul made him at one with sea creatures, but it was less suitable for people. Years later, perhaps out of the water too long, Jacques committed suicide.

His spiritual heir is Francisco Ferreras, better known as 'Pipin', a Cuban living in Miami, where he and his French wife, Audrey Mestre, were routinely breaking all the free diving records. Like Mayol, they were able to reach incredible depths by riding a weighted sled down a steel cable, then inflating a balloon to rocket them back to the surface. Doctors had once predicted that human lungs would be incapable of withstanding pressures over five atmospheres, which corresponds to a depth of 49 metres. Beyond that, they said, the lungs would be the size of potatoes, squeezed so tightly that their tissue fluid would seep out like water wrung from a sponge. In January 2000, Pipin set a new record of 162 metres. The doctors need to recheck their numbers. Perhaps adult humans do retain some vestige of the mammalian dive reflex – what Jacques Mayol meant when he said that while free diving he could feel his acquatic origins.

Breath-hold diving remains very dangerous, however. Even if the human body can withstand the pressure, sometimes technology cannot. Mestre, a twenty-eight-year-old marine physiologist, already held the women's depth record but set out, with her husband's help, to break his record as well. On the morning of 12 October 2002, the weather was stormy off the coast of the Dominican Republic, but by early afternoon the sea had calmed enough to proceed with the highly publicized dive. Surrounded by boatloads of film-makers, reporters and spectators, Audrey Mestre slipped into a meditative trance, detaching herself from the crew on her catamaran who were checking last-minute details. Pipin himself tested the air tank that would be used to inflate the lift bag. He didn't use a pressure gauge but cracked open the valve and heard the reassuring high-pressure hiss. Audrey donned her yellow and black wet suit and slid into the water, which was still slightly choppy from the morning storm. She mounted the metal sled – a vertical tube with two crossbars. Attached to the sled were the compressed air tank, lift bag and a camera with newly added stabilizer wings. Through the vertical tube ran a vinyl-coated cable, held taut by a concrete weight far below. Cables on previous dives had been weighted with lead.

Audrey grasped the top bar of the sled with both hands and folded her knees around the bottom bar so that her fins pointed up. She wore no mask. She took rhythmic breaths until she was ready, then slowly and deeply inhaled the one breath she would carry down with her. She gave a quick nod to release the catch, launching the weighted sled downward. Audrey then plummeted to the world record depth of 171 metres in one minute and forty-two seconds. Waiting there was safety scuba diver Pascal Bernabe, who watched as Audrey uncoupled the ascent portion of the sled and opened the compressed air tank that would fill the lift bag. It didn't fill. Audrey tried a second time. It didn't fill. Thirty seconds had elapsed. Pascal placed his regulator under the bag and pressed the purge valve to force air into it. It was enough to get the sled to rise – but too slowly. Pascal got below the sled and started pushing it up. It repeatedly caught against the cable and several times almost stopped. Audrey remained calm, but she had already been holding her breath over three minutes. The pair moved up to 120 metres before Audrey collapsed on to Pascal like a falling leaf. He continued to swim upward with her until he was met by Pipin, who had donned scuba gear and dived in as soon as he realized she had been down too long. In a grim relay, Pascal passed to Pipin his unconscious wife. They reached the surface far too late for Audrey. It had been 8 minutes and 38 seconds since she took what proved to be her last breath.

What caused the tragic end of Audrey Mestre? The final report concluded that there were many contributing factors. The lift bag did not inflate adequately, possibly due to the compressed air tank not being totally filled. The ascent was slowed because the concrete weight at the bottom was not heavy enough to keep the line taut and vertical, and also because the new stabilizer wings on the camera were pushing sideways against the cable, acting like a brake. The interruptions when the sled actually stopped may have been due to intermittent slack in the ascent line created by ocean swells left over from the morning storm. Audrey's cause of death was listed as accidental drowning. But she wasn't 171 metres under the sea by accident. She had ridden there on the edge of technology – a technology that placed her body at the extreme limit of survival, and then proved too frail to bring her back.

Breath-hold diving tests human limits, but it's no way to explore a vast, unknown environment, which is why I was in full scuba gear as I descended along an anchor line 38 metres to the sea floor, 2 km off the coast of Monaco. That there could be unknown territory so close to the bustling, glamorous city of Monte Carlo gives some idea of how little of the sea has been explored. Less than 1 per cent of the world's ocean bottom has even been seen. The focus of attention of the team I was diving with on that day was a 2-by-1-metre crack running along the base of a submerged limestone cliff. Out of that crack, fresh water gushed into the salty sea at a rate of well over 455 litres a second. Local fishermen have known of the existence of the underwater mineral spring for centuries. From the surface its location was obvious even to me as we approached the site in our research vessel. A patch of smooth surface water about 3 metres in diameter suddenly interrupted the regular pattern of waves. A crewman lowered an empty bottle, brought it up filled and handed it to me to drink: cool fresh water.

I had been invited here to be a medical consultant for NympheaWater, a French undersea research and engineering company about to test the feasibility of capturing that fresh water before it became diluted by the sea, and then providing it as a renewable source of municipal drinking water – in this case to Ventimiglia, a thirsty city in northern Italy. Once they perfect the technology, the company has contracts to set up similar systems in arid countries such as Israel, Morocco, Saudi Arabia and Qatar, underwater springs being plentiful throughout the Mediterranean basin and Persian Gulf. Company president Pierre Becker got the idea while working on another project off the coast of Greece. At a cove near Port d'Itea, he saw shepherds lead their goats a few yards into the sea to quench their thirst. Pierre realized these springs were hidden treasures and he became an underwater prospector. He talks to local fishermen everywhere, studies geological maps, then uses infrared cameras mounted in ultralight aircraft to search the seas for the telltale water-temperature differences that reveal likely sites.

Underground springs are common in limestone cliffs. When water infiltrates and reacts with calcium in the limestone, the acid solution that is formed cuts channels and carves spaces deep within the rock. Rainwater falls into the channels to form rivers and collects in the hollow spaces to form lakes. The network continues its downward flow, emptying out at the base of the cliff. During the last Ice Age, as water became locked in glaciers, the underground rivers and lakes dried up, becoming tunnels and caves. Then, about twenty-one thousand years ago, the climate started to warm up again. Melting glaciers resupplied water to underground rivers and raised the level of the Mediterranean Sea by 183 metres. The outlets of the rivers became submerged. Once the pressure of overlying seawater rises enough to overcome the force of a river, the flow backs up and the tunnel floods with seawater. Should the river flow remain strong enough, however, it will continue to empty into the sea. Being less mineral-laden than salt water, it floats to the surface, where it has long been tempting seafarers, whether on oceanographic research vessels or ancient wooden cargo ships. Three thousand years before Pierre Becker, Phoenician sailors covered undersea mineral springs with bronze bells and brought the water to the surface through leather tubes covered with tar for waterproofing. The Nymphea Water project is simply an updated version of their extraction system.

On the sea floor, I held on to a rock outcropping with one hand while a sensor in my other hand measured the temperature and salinity of the upwelling water. Divers around me were placing concrete blocks in a circle, and divers above me were deploying the collecting bell – a blue canvas sheath that looked like a closed umbrella with dangling ropes. One by one each rope was attached to a ring in one of the concrete blocks;

then a long supple tube, a chimney with a wide opening at the bottom, was placed over the mouth of the spring. The water rose naturally through the tube, snapping it straight before exiting through the opening at the top, directly under the dome of the bell. The fresh water was trapped inside and filled the bell like a hot air balloon. The bell expanded into a gossamer dome 2 metres across and 1 metre high, pale blue against the deep blue dimensionless background. As the fresh water bubbled in at its centre and the salt water that it replaced spilled out around its edges, I found it hard not to imagine that the bell was a breathing organism: a giant blue jellyfish with rope tentacles, hovering just above me. I was mesmerized. The longer I watched, the more I believed it was alive. I knew I had to move, but not because I thought it would attack. It wasn't frightening – after all, the jellyfish's tentacles were tied into concrete blocks – but the image was too real. I had been too deep too long. I needed to ascend a few dozen feet to clear my head and reverse the early symptoms of nitrogen narcosis – rapture of the deep.

Breathing at depth is possible only because compressed air is delivered at the same pressure as the surrounding water. The scuba regulator senses the outside pressure and opens the air valve to match it exactly so that a diver's lungs inflate with no resistance. The simple system has allowed divers to adapt to otherwise inhuman depths. However, forcing air into the body under pressure creates its own unnatural and very strange problems. The first human bodies to react to the effects of breathing air under high pressure belonged not to divers but to tunnel and bridge workers. To keep water out of their workplace, compressed air was pumped into partially completed tunnels, or into wooden tubes, called caissons, lowered into rivers where workers were constructing bridge pilings. Often, by the end of the day, the mood of the entire crew turned euphoric, with workers singing and laughing spontaneously. The cause of the strange behaviour remained a mystery for many years because the guilty agent had been almost above suspicion. Nitrogen is an inert gas that makes up nearly 80 per cent of the air in our atmosphere. Humans have been breathing it as long as they have been human – but always at atmospheric pressure. When nitrogen is inhaled under increased pressure, it becomes a narcotic.

Humans have no use for nitrogen gas, which passes in and out of the lungs during respiration without the body even noticing. None is absorbed, because gases, like nitrogen, do not mix readily with liquids, like body tissues. Gases and liquids can be made to mix under pressure, however, as is done when adding carbon dioxide to beverages. Breathing compressed air is like getting carbonated. Nitrogen gas is forced into the blood, then picked up by the liquid in individual cells, especially in the brain. Once inside, it interferes with the transmission of electrical impulses, although how this happens is only poorly understood. The effect is predictable enough, however, for doctors to rely on certain compounds of nitrogen to provide anaesthesia. Nitrogen also blocks inhibitory pathways that moderate behaviour, in much the same way alcohol does, which is why divers who have gone too deep begin to feel drunk or hallucinate.

Before the giant blue jellyfish could grab me, I ascended about 9 metres, to a depth that would depressurize the nitrogen I was breathing. Within a few minutes the gigantic sea creature had turned back into a canvas bell. The scene was nonetheless captivating. The system was working. Suspended in the sea, we were collecting water from a spring that once flowed on dry land 91 metres above sea level. Men of the Ice Age must have come here to drink from this powerful source of fresh water. Less powerful flows would have dried up, leaving behind tunnels that led to caves deep inside the cliffs. Caves that Henri Cosquer's ancestors could have walked into.

The NympheaWater team was ready to celebrate the success of its pilot project, but had we left the water quickly, we would all most likely have been dead in a few hours. This environment, which requires such meticulous planning to enter, demands even greater care and caution to leave. A quick exit would bring on the bends, the other disease brought on by breathing air under pressure. It too was first described in tunnel workers, and was originally called caisson's disease, but unlike nitrogen narcosis, whose onset occurs when a diver goes down, the bends doesn't strike until a diver tries to come up. Depth, time and exertion all contribute to the amount of gas that is forced into the tissues. This dive had a lot of all three – more than enough to turn the liquid portions of our bodies into carbonated beverages. Had we surfaced abruptly, the sudden release of pressure on us would have the same effect as popping the cap off a soda bottle. We would have fizzed up.

Dissolved gases that come rapidly out of solution form bubbles. When it happens inside the body, the dissolved nitrogen that has entered quietly bursts into a rampaging enemy that is everywhere at once. Blood vessels that had unwittingly transported hazardous material are suddenly filled with tiny air bubbles rapidly coalescing into large ones. The pale yellow lubricating fluid in joints now looks like sparkling champagne. Inside individual cells all over the body, pressure rises as bubbles form within their liquid milieu and expand against their cell membranes. The alien invader is not like anything the body has ever confronted in its entire history of evolution, and consequently it has never developed any defence for it. Its response is as chaotic as earth's would be to an invasion from Mars.

Having gained access to the body's internal transportation system, the bubbles are carried by circulating blood deep into organs and tissues, stopped only when vessels become too small for them to pass through. At that point they occlude the channel, preventing any further flow of blood. Without their supply of oxygen and nutrients, tissues and organs downstream of the blockage starve. Within the blood itself there is mass confusion. White cells, which travel in the blood patrolling for infection, encounter the invader and interpret it as a new form of bacteria. They react by creating an inflammatory response, as if fighting an infection. Platelets, travelling blood cells that institute clotting, sense the presence of air. For them, that means that the blood vessel has been cut and there is a breach from outside. They activate the clotting mechanism to stop bleeding that isn't occurring, forming blood clots that have nowhere to go. Outside the blood, things are no better. Bubbles squeeze the contents of individual cells beyond the cell membrane's ability to contain them – cells throughout the body randomly burst.

The only way to prevent this anatomic catastrophe is to let the pressure dissipate very gradually. A soda bottle can be opened without much fizz if the cap is released gently and a little at a time. A diver can emerge unscathed from the depths if he ascends slowly and stops at prescribed intervals. That's why our team members were moving toward the surface little by little, stopping at various points along the anchor line or along safety lines dropped from buoys. Air tanks had been tied to the lines at specific depths to provide decompression stations where we would have the extra oxygen we needed to extend our time underwater. The number of stops required, the depth at which to make them and the length of time at each one has been worked out in decompression tables drawn up by the US Navy. The schedule depends on the diver's maximum depth and time underwater. Each of us was following his or her individual prescription to avoid the bends.

Whenever I'm tempted to take a short cut to the surface, I think about my longtime friend and dive buddy Bernie Chowdhury, who got severely bent a few years ago, but lived to tell about it. Bernie is the publisher of *Immersed*, a high-end scuba magazine, whom I met when he invited me to write an article. He's also a consummate diver. One day when he was exploring the wreck of an ocean liner off the coast of Delaware, however, he was pushed to the human limit. He recounts the story as a part of his book, *The Last Dive*, and I've heard him retell it many times. Each time I hear it, I can't help wondering whether Bernie didn't actually exceed the human limit – and still make it back.

The wreck in question, the *Northern Pacific*, was stuck bow-down in the sand at a depth of 46 metres. It was Bernie's third dive to it. He wasn't feeling well that morning, and being a cautious diver he would ordinarily have followed his instincts by not diving. However, he had already gone down twice and decided there wouldn't be much risk in diving it again. When he arrived at the upended stern he tied off the extra air tank he would need later for his decompression. Then he swam down the 152 metre length of the ship and entered the bow section, looking for portholes, dishes and other artefacts, or at least some lobsters. He didn't find any souvenirs, and after bagging his second lobster, his dive time was up. He exited the bow, intending to swim back to his decompression tank tied to the ascent line at the stern. The current had picked up, though, forcing Bernie to kick harder and harder to make headway. To get out of the current, he dropped below the ship, to the sandy bottom, where he made better progress. But the increase in depth, combined with all the exertion, was enough to give him a dose of nitrogen narcosis. Bernie says he felt as if he had instantly gulped a martini. His head wasn't clear – and neither was the water so close to the seafloor. When he reached the stern he couldn't find the anchor line. He kept swimming around, looking frantically for his spare air tank, until he recognized with a shock that he was back at the bow of the ship. He had somehow swum underneath the raised stern and circled completely around the boat. Bernie weighed the odds of making it back to the stern and finding the spare tank before his dwindling air supply ran out. He still had enough air to ascend, but only if he didn't stop to decompress. He was physically and mentally exhausted. The stern was 152 metres away, and he couldn't see more than 9 metres through the cloudy water. The odds against finding the spare tank were too high. Bernie decided to go straight to the surface.

The pain started while he was bobbing in the water, his friends trying to pull him into the dive boat. There was some thought of taking Bernie back down with fresh tanks and then bringing him back up slowly – an in-water recompression – but his pain became so intense so quickly that he would have drowned had he stayed in the water any longer. He was hauled out, placed on the gear table and given oxygen. He was having trouble breathing, and the pain was everywhere; he began seeing in triplicate; he lost hearing in both ears; strength and feeling were slowly leaving his arms and legs; his pulse was steadily weakening. He realized he was facing death.

Bernie writes, 'The pain eased and a feeling of well-being washed over me. A bright white light appeared . . . my body drifted upward . . . I could see everyone aboard and my sorry self lying on the table. I went into a white tunnel. I had only to float to the end and all my problems would be over. I drifted closer to the end of the tunnel and the light got even brighter.' Then Bernie remembered his wife and son and how much they needed him. 'I stopped my drift and struggled to turn around. I had to get back. The white light dimmed and then went away completely. I opened my eyes. Searing pain racked my body again.'

The impression of floating over the scene, the out-of-body experience, has often been described by people who have seemingly died and then come back to life. It's reminiscent of Pablo Valencia lost in the desert and Beck Weathers, an Everest climber whom I treated after he collapsed in the snow. It's also reminiscent of my jungle friend Antonio describing the effect of his hallucinogenic drug. What these experiences have in common appears to be that when body function shuts down, the highest cerebral centres remain active and, dissociated from any outside input, create from within a vibrant image free of any space-time underpinning. The image has electrical power and if intense enough can restart activity in the rest of the body.

Bernie willed himself back to the dive boat. He focused on surviving one step at a time. First he had to last until the evacuation helicopter arrived. Then he had to endure the flight to the recompression chamber. Maybe he would start to feel better after recompression, but maybe not. He didn't know how much of his body had been permanently destroyed. He was having difficulty breathing because bubbles had lodged in the small capillaries surrounding his lungs, blocking air exchange. He couldn't walk, because bubbles had entered his joint spaces. He couldn't hear, because both inner ears had taken direct hits. Most ominous were the nerve problems. Parts of the fine plexus of blood vessels around the spinal cord had been occluded, damaging the nerves they supply by depriving them of oxygen. With signal transmission pathways interrupted, Bernie was losing control of his voluntary muscles and losing awareness of his skin sensation. Nerve cells in various parts of his brain were malfunctioning, leaving him unable to balance or focus both eyes simultaneously. He was somewhat confused, but his mind was still clear enough to comprehend his predicament and to fear death, a healthy fear that no doubt reinforced his determination to stay among the living.

Forty-five minutes after he surfaced, Bernie was lifted into a Coast Guard helicopter and flown to a recompression chamber at the Hospital of the University of Pennsylvania. The principle of recompression is to reapply the maximum pressure to which the body was subjected, so that the gas bubbles go back into solution, then to reduce the pressure gradually enough to prevent them from reforming. Bernie spent six hours in the chamber. When he came out he was in less pain, and his vision and hearing had improved, but he still had no balance, couldn't walk and had difficulty concentrating. Some of his nerve tissue had been permanently destroyed, and his brain and body had to adapt to new ways of doing things. It took over a year, yet somehow – incredibly – Bernie made a full recovery. He is even diving again.

Even when they have happy endings, stories such as these offer powerful reminders not to skimp on decompression time. Forty minutes after I left the bottom, I was still hanging on the line, looking up at the hull of the research boat, only 3 metres above me. Finally my calculated decompression time, with a safety margin added, was completed. But I still wasn't free to just swim up to the boat. There was another invisible barrier between the surface and me, one I had to cross carefully if I wanted to leave the water alive.

From my maximum depth of 38 metres, I was able to ascend 19 metres before the pressure on my lungs was halved (discounting overlying air pressure). From the 3-metre depth where I was now, the pressure would halve with a further ascent of merely 2 metres. But the rapid decrease in pressure as I move toward the surface will mean a reciprocally rapid expansion of the volume of gas in my lungs. That air needs time to be either absorbed or exhaled. If expanding air remains trapped inside the lungs, a pressure change of only 0.9 or 1.2 metres is enough for the air to overcome the elastic limit of the alveoli, rupturing the delicate membranes and sending air directly into the pulmonary circulation. From there it usually flows up the carotid arteries in the neck and lodges in the brain, creating the same symptoms as a stroke.

The only safe way to ascend is to exhale continuously and to stay below your smallest bubbles. The bubbles will rise slowly enough to give the excess air in your lungs time to escape. However, a diver running out of air or in a panic may hold his breath and try to get to the surface as quickly as possible. Betrayed by his survival instinct, his lungs will burst as soon as the water gets shallow. The last 3 metres of the undersea world is a treacherous border.

Finally we crossed safely and were able to celebrate our success in true French style – champagne all around. The system had worked, everyone was back on board and the only bubbles were in the drinking glasses. Had the spring been a little deeper down, we would all still be on the ascent lines – at least those of us who hadn't succumbed to nitrogen narcosis and been lured away by mermaids. Decompression time increases rapidly with depth, as does the risk of rapture of the deep. Ascent time soon exceeds bottom time, thinking can become cloudy and you need to save most of your air supply to get back up. The relentless pressure of nitrogen in compressed air creates invisible but very real barriers to penetration of the deep sea.

Given that nitrogen is an inert gas, serving no purpose in the human body and causing real problems under pressure, why not simply take it out of the compressed air? The problem is that it has to be replaced with something, explained my colleague and dive physiology mentor, Dr Bernard Gardette of Comex. The world's foremost underwater engineering and construction company, Comex runs operations in every ocean in the world, but their world headquarters is in Marseilles, France, just a few hours by boat from where we were diving. The company's founder and chairman, Henri Delauze, had lent Pierre Becker his state-of-the-art research vessel and equipment. And he had lent me Dr Gardette, his chief medical officer.

The answer to the nitrogen problem that had immediately occurred to me was to breathe pure oxygen. I and most other climbers do that routinely high in the Himalayas because it eases breathing in a low-air-pressure environment. The sea, on the other hand, is a high-water-pressure environment, and concentrated oxygen under pressure turns toxic. It is readily metabolized and therefore would not cause the bends, but when absorbed in high doses it irritates nerves and can cause convulsions and sudden loss of consciousness – not a good combination, especially underwater.

No, the oxygen must remain diluted, so nitrogen must be exchanged for some other inert gas – helium, Dr Gardette informed me. Mixed with oxygen, and called heliox, helium immediately eliminates the problem of narcosis. Since helium is far less soluble in water than nitrogen, the liquid human body can withstand far more pressure before it absorbs the gas. Helium is a very light gas – light enough to float a dirigible – so it remains easy to breathe at depths where all gases become densely compressed. It does have disadvantages, however. Divers get cold breathing helium because it is too light to hold much heat. Also, its decreased density makes vocal cords vibrate faster. Divers on helium talk like chipmunks. It's not funny – they can't understand one another and need computerized unscramblers to communicate.

Nonetheless, heliox has pushed the undersea frontier down below 244 metres. Deeper than that, divers become prone to a bizarre new malady: high-pressure nervous syndrome, in which divers become giddy and shaky, apparently due to helium's finally being absorbed, or perhaps due to sheer pressure. Ironically, when nitrogen is added back into the divers' breathing mixture, its narcotic effect calms them. Even helium gets heavy at these enormous depths, particularly when combined with nitrogen. To get humans deeper, Dr Gardette and others have been experimenting with hydrogen, a gas that weighs only half as much as helium but is highly explosive. Handling it requires extreme precaution, as I saw when I toured the Comex facility, replete with gas sensors, huge ventilation ducts and *No Smoking* signs everywhere. But hydrogen is now routinely handled as fuel in space exploration; certainly it can also be used safely as a breathing gas to explore the ocean depths.

We may never know how much deeper hydrogen might take us, however. Human penetration to ever greater depths has been stopped by a barrier perhaps more formidable than anything that would have been encountered in that hostile environment – marketplace economics. With the advent of unmanned deep-sea submersibles that can descend hundreds of metres with TV cameras and robot arms, it has become easier and cheaper to let a robot do the job. The need to send humans to those extreme depths has become obsolete.

Nonetheless they are still very much needed at 'moderate' depths, especially for work involving offshore oil rigs and pipelines. People all over the world live and labour in oceans at depths of 305 metres or more. They are blue-collar workers – welders, mechanics, electricians – who go to their jobs in diving bells, stay a month or so and then return to the surface after a weeklong commute. To be sure, these people are also highly trained divers, specializing in the art and science of saturation diving.

Using standard dive technique below about 50 metres, bottom time becomes so short and decompression time so long that the sea becomes an impractical place to get any work done. Commercial divers need a better way to get to work, and they need the time to get their jobs done right without having to watch the clock to see how much gas they're absorbing. Fortunately, there's a limit to how much gas a liquid can absorb. Human bodies become maximally saturated after about twenty-four hours at any given depth. After that, no matter how much longer you stay, you don't absorb any more gas. That means your decompression 'obligation' doesn't increase. Once you're down that long, you may as well stay awhile.

A thick glass porthole was all that separated me from the saturation diver I met at sea. I was breathing fresh salt air. He was breathing heliox, and his body was under 183 metres of pressure. He smiled, said, 'Bonjour', then turned back to his card game with his dive buddy. They were in the hull of a ship inside a hyperbaric chamber, a thick metal canister containing a living room, bedroom, bathroom and a lock-in/lock-out chamber that overlooked the moon pool – a hole in the bottom of the ship. When they started the job, they entered the chamber at atmospheric pressure. The chamber was then gradually pressurized until it equalled the depth of their intended work site. The two divers donned their elaborate gear with the help of a bellman, a combination diving bell operator, troubleshooter and valet. All three passed through a hatch into the diving bell, which is hermetically clamped to the lock-out chamber and hangs above the moon pool. The diving bell was lowered deep into the dark cold sea, tethered by a cable that contains compressed gas hoses, electrical power and telecommunications wires, and tubes to circulate hot water. When it reached the designated depth, the pressure inside and outside the bell was equal. The door at the bottom opened outward easily. The divers swam out to their workplace, each tethered to the bell by a second cable, a lifeline that provided breathing gas, light, heat, two-way communication and biomonitoring. The bellman stayed behind in the bell, monitoring the support systems and keeping a watchful eye on his two charges. He continually adjusted the length of their umbilical cords to allow the divers to move freely yet without slack, which might entangle them. As an extra measure of security, each diver carried a bail-out bottle with a small supply of breathing gas in case his lifeline was accidentally cut or in case he became entrapped and had to cut it himself. The extra minutes would give the diver enough time to return to the bell – his only possible refuge.

Emergencies are, surprisingly, very rare. With meticulous preparation and intense supervision of their habitat, commercial deep-sea divers have far fewer accidents than do even the most careful shallow-water divers. They must, however, remain completely enclosed in their environment; rapid depressurization due to a break in any of the seals in their system would immediately inflict the bends. A doctor has access to them through the lock-in hatch of the hyperbaric chamber, but for a major medical problem, or a fire or other emergency, the divers cannot get out quickly. Evacuation means transferring them to a portable capsule that can be loaded on to, and hermetically linked to, a helicopter equipped with a hyperbaric chamber. Then they have to be offloaded, still under pressure, to the medical recompression facility for treatment.

However, very few saturation divers leave the work site that way. Statistically the profession is no more dangerous than the same blue-collar occupations are on land. When the day's work is finished, the divers swim back to the bell, take the elevator ride to the surface and enter the hyperbaric chamber, where they shower, rest, relax and remain under high pressure, awaiting their next shift. Going to work at the frontier of the most hostile environment on earth becomes a daily routine. After following the same schedule for several weeks, they stop diving, so they can ready themselves for some R&R. Though only a few metres away from everyone else, they are separated by hundreds of pounds of seawater pressure that has to be reduced very gradually. It's a long, slow, boring trip back to their native habitat. Depending on the depth at which the divers have been working, they may need seven to ten or more days of decompression before they can breathe fresh sea air.

For a large operation, teams of these saturation divers are 'stored' onboard ship at various pressures so that they can be injected into the sea to work simultaneously on parts of the project located at different depths. It's easy to imagine each group of pressurized divers as an artificially created sea species with a habitat limited to a narrow depth range. In a cross-section of the ocean, they would be stacked one below the other, sharing their designated layer of water with the deep-dwelling fish that are confined to the same narrow zone by their millennia-old adaptations.

Most fish have swim bladders, a space within their bodies that controls buoyancy. When a fish secretes gas into the space, it expands and the fish rises. When the fish absorbs the gas back into its blood, the bladder contracts and the fish sinks. It's an effortless way to move vertically through the water. Bladders have limited capacity, however; they have been preset by evolution to function within a depth range specific to each species. Saturation divers have evolved their own version of a swim bladder: the buoyancy-compensation vest – a jacket that can be inflated with air from the scuba tank or deflated by bleeding air out through the regulator.

In the sea, deeper is darker. Many deep-dwelling fish have grotesquely enlarged eyes, and many are bioluminescent, creating their own light using chemical reactions similar to those of fireflies. Humans don't have huge eyes, and their lenses are adapted to focus light rays travelling through air. Because water bends light enough to change the angle at which it hits the lens, underwater images appear blurry to us. To see clearly in the water, humans need to enclose air in front of their eyes by means of a face mask. The mask not only refocuses the image, it intensifies and enlarges it, since light travelling through dense water into thin air is refracted in the same way it would be were it to pass through a magnifying glass. Images appear closer, bigger and brighter. When there is no light at all, a diver will bring along headlamps and flashlights. He might also carry a light stick, a clear plastic tube containing two liquids separated by a delicate partition. When that partition is broken – which the diver can do by snapping the tube – the liquids mix to create a bioluminescent glow.

There are some adaptations a human cannot mimic. Every living cell of every living creature is enclosed and protected by a cell membrane, a somewhat flexible covering of protein and lipids. In humans, the lipid is a saturated oil that, like butter, will solidify at cold temperatures and deform under pressure. The cell membranes of deep-sea fish are made of unsaturated oils that, like olive oil, remain liquid at low temperatures and can therefore withstand the cold and high pressure of the sea much better.

The most fundamental difference between deep-sea fish and humans, however, is also the subtlest. Ongoing biochemical processes in living organisms create constant changes in the three-dimensional shapes of proteins and in the total volume of molecules produced by chemical reactions. When they are placed under intense pressure, proteins are not able to unfold, and expansive reactions are constrained. Deep-sea fish have undergone evolutionary modifications in their protein design and in their metabolism to allow protein unfolding and energy production to occur without any increase in volume. In humans at the same depths, those chemical reactions would be smothered by the surrounding pressure.

Deep-sea fish are so exquisitely adapted to their extreme environment that they are unable to survive when brought too near the surface, where they are exposed to the unhealthy conditions of light, warmth and low pressure. They won't get the bends, since their gills have been extracting only oxygen from the water they breathe. Their liquid-oil cell membranes begin to ooze, however, responding to the lack of any containing pressure. Some fish literally fall apart as their chemistry goes haywire. Others explode when their swim bladders overexpand and burst, like the lungs of a diver who ascends while holding his breath. Actually, very little is known about these denizens of the deep. They are extraordinarily difficult to collect, and they self-destruct before they can be studied.

We not only know very little about life in the sea, we know very little about the sea itself. Huge currents flow between and around the continents, currents of which we have only the vaguest awareness, even though they largely determine our climate and weather patterns. The most studied of those currents, the Gulf Stream, brings heated equatorial water northward across the Atlantic, where it warms the European continent. How effectively it does that is readily apparent when you consider that the French Riviera and Nova Scotia, Canada, are at the same latitude. In the 1970s, an enormous lake of cool fresh water appeared just below the surface of the North Atlantic, probably the result of an ice melt off the coast of Greenland. The water chilled the Gulf Stream for several years – years that corresponded to some of the coldest winters Europe has known. When the lake finally dissipated, the balmier weather returned.

In a highly ambitious attempt to learn more about what is going on in the sea – and who and what might be living there – plans are under way to construct an international subsurface research station that will drift around the world for two years on the great ocean currents. The 37-metre tall aluminium station, named *SeaOrbiter*, will resemble a boat bobbing vertically in the water. The upper half will provide a platform for atmospheric measurements and satellite communications; the lower half will constitute a continuous undersea laboratory for studying the biology, chemistry and physics of the ocean. The station projects 15 metres below the surface, with a panoramic window near the bottom from which to observe strange sea life – not the least of which will be the saturation divers who will come and go from the hyperbaric chamber on the lowest deck. Six divers will remain permanently in saturation for each leg of the voyage, giving them quick, easy and anytime access to the environment while also, not incidentally, providing models for studying the effects of prolonged saturation.

As a member of *SeaOrbiter*'s scientific committee, I have discussed the project at length with its chief architect, Jacques Rougerie, and met with many of the researchers and engineers. As a member of *SeaOrbiter*'s crew when it is cast adrift, I will be at one frontier of the sea, helping to probe its deepest mysteries. Even at its deepest point, the ocean's bottom will always be less than 11 km below our research station. Yet the sea remains the least explored, least understood environment on earth because it is the most hostile to man. The environment from which all life arose and on which all life still depends has become, to humans, through the nearly imperceptible small steps of evolution, the most extreme environment on earth.

HIGH ALTITUDE

IN THE KINGDOM OF THE GODS

HALF A BOWL OF WARM WATER is not much to eat, but at midnight in a tent at 7,925 metres, I was happy to have even that. One of three climbers crammed into a two-man tent, I was trying to rest but the temperature inside the tent was –18°C, and it was frightening to think how much colder it must be outside. The weather was holding, though; we couldn't let a good chance like this go by. After two months in this hostile environment, we were dehydrated, malnourished, oxygen-starved and exhausted, but if our bodies could hold out another sixteen to twenty hours, we could reach the highest point on earth.

There were nine of us in the three tents we pitched on the South Col, 914 metres below Mount Everest's 8,850 metre summit. These altitudes are the domains of jumbo jets cruising high over the oceans, not creatures born and raised near sea level. Our bodies were deteriorating rapidly, consuming themselves to produce enough energy to function in a world with hardly any heat or air pressure. Though our expedition had arrived at this camp only six hours before, after days of relentless climbing, it was already time to move on. Too prolonged a stay could become a permanent one.

I took off my oxygen mask, disconnecting myself from the bulky tank lying on the floor, unzipped the door flap and crawled out of my thin nylon shelter. The sky was black except for a sharp silver moon, two-thirds full, beaming just enough light to make the ice all around us glow deep blue. The still air was actually not much colder than it had been inside the tent. So effective were our insulated clothing and sleeping bags at retaining body heat that little had escaped to warm the interior of the tent. Insulation only conserves heat, however; it doesn't produce it. To do that, the body needs to burn fuel. Burning fuel requires oxygen, and now that I was separated from my supplemental supply, my body would have to make do with what it found in the air around it, which is about one-third the amount at sea level.

My breathing turned deeper and more rapid and my pulse quickened as I put on my climbing gear. Some of this was due to nerves – I was about to attempt the summit of Mount Everest – but most of it was my body's automatic response to the lack of oxygen. My lungs were working harder to drive more air inside, and my heart was beating faster to pump the limited oxygen supply to where it was most needed. Even so, I was feeling much colder by the time I had buckled my harness and hoisted my pack. Though the outside temperature hadn't changed, my body's ability to cope with it had. Without the extra oxygen, my internal combustion was slowing. Fuel can't provide energy when there isn't enough oxygen to burn it. My metabolism was becoming a smouldering fire too weak to generate enough heat to keep me warm.

A group of fragile humans were about to set off on a journey through a dark, frozen, air-depleted and lifeless portion of the planet, our survival dependent upon the life-support system we were bringing with us: lights, clothes, food, water and oxygen. We would be under attack every step of the way. All we could do was hope that our portable mini-environment would keep our bodies from realizing where they really were.

The first line of defence is clothing. Mine consisted of three layers: next to my skin, polypropylene underwear – to keep me warm while wicking away the sweat that would otherwise accumulate on my skin and dissipate my body heat thirty times faster than if my skin were dry; next, a one-piece suit made of down – a goose's soft under-feathers designed by nature into a pattern that creates millions of tiny heat-retaining air pockets;

my outer layer was a nylon shell – to act as a windbreaker and to keep my inner layers dry despite constant contact with ice and snow.

Heat loss is greatest at the body's 'ends' – head, hands and feet – where surface area is large in comparison with the volume of tissue inside. I wore a tight-knit wool hat with a balaclava tucked inside – a wool ski mask that could be folded down to cover my face and neck, leaving only slits for my eyes, nose and mouth. My eyes were covered by goggles to block the wind and the ultraviolet radiation, my nose and mouth covered by an oxygen mask.

Keeping hands warm remains an unsolved problem in mountaineering. Fingers, being long thin appendages, are very vulnerable to freezing. Protecting them requires heavy insulation, but gloves frustrate the dexterity a climber needs. No glove will keep a hand warm if it isn't worn. The compromise usually consists of two pairs of gloves: a thin, conforming inner pair covered by an outer bulkier mitten – not having finger separations, mittens have greatly reduced surface area and thus retain much more warmth. For tasks that require individual fingers, the mittens are removed and left dangling from little strings attached to the jacket sleeves (the way toddlers wear them) so they won't blow away in the wind. It sounds fine in theory, but in practise mittens are annoying, and they are off far more often than they should be.

Feet are easier to insulate than hands because toes are stubbier than fingers and don't have to move independently. Two pairs of socks – one polypropylene, one wool – enclosed in plastic double boots can amply protect them. A metal platform with spikes – a crampon – is attached to the sole of each boot to grip the ice. Though metal is a rapid conductor of heat, and the crampon is in nearly constant contact with snow or ice, the double boot, with its layer of trapped air, acts like a thermos to keep body heat inside.

Even perfect insulation won't protect a body that's not generating heat. As we stood around checking our gear, our own as well as each other's, then waited for everyone to make last-minute adjustments to satisfy themselves that they were absolutely ready, the relative inactivity caused me to shiver – not in anticipation of what lay ahead, but because my body temperature had dropped far enough to alarm the thermostat in my hypothalamus. It signalled my muscles to shake in order to produce heat. My body was stoking its internal fire.

The fire would burn a lot hotter once we were under way. The large muscles in my arms and legs would begin working, providing the heat to rewarm my body as well as the power to carry me, I hoped, to the summit and then back. The round-trip, however, would require a great deal of oxygen, fuel and water, all of which were in limited supply. My backpack contained two lightweight oxygen cylinders and a few chocolate bars. A bottle of hot water was tucked inside my down suit to keep it warm. We had to be self-contained and self-sufficient. There would be no food and, since we had neither the fuel nor the time to melt snow, no water. Most dangerous of all, there would be not nearly enough air.

I turned on my regulator, set it for a rate of 2 litres per minute and made sure there were no kinks in the hose that directed the oxygen flow from the backpack and wound under my arm and up to my nose and mouth. I slipped the mask over my head and cinched the straps to create an airtight seal against my face. After a few reassuring breaths of oxygen, I lowered my goggles into place, turned on my headlight and took my first steps toward the summit of Mount Everest.

We left Camp IV, the encampment on the South Col, with high hopes and strong determination. None of us would have gotten this far without plenty of both. Sheer will can take you deep into an inhospitable environment, but you still have to bring your body along, and though the day was just starting, our bodies were already fast approaching their limits. We had been moving up and down the mountain for two months, burning vast quantities of energy without replenishing the fuel supply adequately. High altitude blunts sensations of hunger and thirst. If things went well, I would need to burn 12,000 to 15,000 calories on this day – ten times what the body uses in an average day – and maybe more than I had in reserve.

The route to Everest's summit starts out across the Col, a broad, flat ice sheet that ends in a snow gully. In the darkness, our expedition became a procession of lights, each climber a yellow dot against the deep blue of the ice. The air was cold but my high rate of activity, sustained by the steady flow of oxygen, was enough to warm my body. No wind broke the silence. I listened to the rhythm of ice crunching under my crampons and oxygen flowing in and out of my lungs.

We entered the snow gully, which first gradually and then quickly steepened into what is known as the Triangular Face. Our line of lights became more vertical, the spacing more irregular as each climber progressed at his or her own rate. As I moved up the Face, the headlight in front of me came steadily nearer. The climber seemed to be losing power early. Finally, his headlight stopped. My light soon illuminated the back of a heavy down jacket – and the cause of the climber's exhaustion. He was overdressed. He had been working strenuously enough to overheat despite an outside temperature lower than -18°C.

The diagnosis was easy – though heat exhaustion is not one of the illnesses you think of first when you consider the risks of high-altitude climbing. The treatment was easy too: take his jacket off and let the cold air catch up with him. But nothing is simple near the summit of Mount Everest. The snow we had been climbing through was too powdery to kick steps into, and we had been relying primarily on our hands for support, not trusting our footholds. To free our hands in order to remove the climber's jacket, we would have to shift all our weight to our feet, and we weren't sure the snow would hold. Other climbers came up behind me, and we worked out a treatment plan. I was the expedition's doctor but far from its most skilled climber. I had prescribed the appropriate treatment based upon my diagnosis, but that treatment would be carried out by two of my stronger teammates as I moved up the slope a bit more to watch. They placed an ice screw in the snow above them, fed a rope through it, then hooked it to their harnesses. Hanging from the rope – to which they also attached the heat-exhausted climber – they could keep their hands free to work off his jacket. Once it was removed, the climber revived dramatically in the cold air, so much so that he soon passed me as we all continued upward.

With the delay, the line of lights was now fairly high above me. I tried to quicken my pace to catch up but didn't have enough oxygen. Reaching a crest in the Triangular Face, I was momentarily relieved to see that the route became less steep and that the others weren't too far ahead. As I crossed the crest, however, I immediately realized why the others hadn't progressed farther than they had: I plunged into much deeper, drier snow. This would require more energy and more time to go less distance as well as more oxygen to sustain the extra effort. With higher steps and heavier breathing, we made our way toward the top of the Triangular Face. Our regulators were set on 'continuous flow', not 'demand flow', and since we had all dialled in the same rate – 2 litres per minute – our first tanks would all be empty at about the same time. Some climbers were already stopping to change to their second tanks. My gauge indicated that my tank was nearly empty, but I wanted to use every ounce of oxygen I had, so I continued on toward a small rock outcropping that formed a flat spot in the slope. Two other climbers were there already, helping each other change tanks as I approached.

Suddenly my body weight doubled, pulling me down into the snow. I could still see the two climbers ahead, but not the rock on which they were standing. A flickering ring of lights obscured all but my central vision. I felt detached from the scene – as if I were no longer a part of it. My reasoning remained intact, though; I knew I was out of oxygen. A teammate came up behind me, unscrewed my regulator and attached it to the second tank. My head cleared up instantly, my peripheral vision returned and strength flowed back into my legs. Oxygen was a wonder drug.

I reached the top of the Triangular Face at the same time as everyone else – an hour after sunrise. The plan had been to arrive there just as the sun was coming up. We had lost an hour of daylight that we should have used climbing the treacherous southeast ridge that leads to the summit. And we had used too much oxygen to get this far. Tanks should have been changed on the ridge, not before. To conserve what we had left, we had to turn our flow rate down to 1.5 litres per minute. With less oxygen, our chemical reactions – needed for thinking, talking, judging – would become less efficient, our bodies would cool, we would move even more slowly. But there was no other choice. It would be far worse, and maybe fatal, to run completely out of oxygen at an even higher altitude.

Only the southeast ridge remained between us and the summit, 457 vertical metres higher. The ridge is a long, narrow, crooked seam of rock angling 30° upward and tilting 15° from left to right. After a few minutes' rest for water and chocolate, we climbed on to the ridge. The world opened below me. Standing on an edge barely 2 metres in width, I looked down on ice-covered mountains. I could see the curvature of the earth. I had climbed into outer space.

Gazing up the ridge toward the summit, we saw more of the same deep snow we had been wading through all night. The ridge would be a lot trickier than the Triangular Face, and we would be doing it with less oxygen. The thought that we were not going to make it was in everyone's mind, though no one said it. We stepped carefully along the ridge in single file, unroped, mindful that at a lower altitude one unroped climber on our expedition had already taken a fatal plunge. But the slope here was so sheer that a slip would turn into a free fall – impossible for another climber to arrest. A rope would just jerk a second climber off the edge when one would be more than enough.

The snow was knee-high at the start of the ridge, but soon we were plowing through waist-high snow so dry and granular that it collapsed back around us as soon as we had passed through, forcing each climber to break his own trail. We were climbing a mountain of sugar. Unable to see my legs, much less my boots, I was walking a crooked line, never quite sure where my feet were. With each step, I tried to feel around with my crampon for firmly packed snow before putting my weight down. After several hours of tentative, measured steps, we had made very little progress. Even more discouraging, the angle of the ridge was increasing, the snow getting deeper. Still, the weather was holding. There was little wind, and except for occasional periods of 'fog', which at this altitude were actually clouds moving past us, visibility was good. I had to wait for the air to clear whenever a cloud came by; otherwise I couldn't be sure if my next step would be on snow or into thin air.

We were at 8,565 metres – at the buttress of the South Summit, a prominence in the ridge after which the angle lessens. The summit was only 274 vertical metres higher. Under normal snow conditions we could be standing at the top in less than two hours. But we were in snow that now had become chest-high. The climb had been gruelling – depleting both our oxygen reserves and our energy reserves. We knew how much oxygen we had left; we could only guess how much energy we had left. It was noon. We calculated that at our current rate of progress we would reach the summit in four or five hours. That would mean, though, that we would have to come down in the dark, most likely exhausted and certainly without oxygen. Decisions aren't easily made by cold, thirsty, very tired climbers whose brains have been short on oxygen for hours. But a decision had to be made.

The last time we had all been at this altitude we had been thinking a lot more clearly. It was when our plane began its descent into the Kingdom of Nepal. Outside the windows, at eye level, was a massive pyramid of rock piercing a carpet of clouds, alone in a vast, empty space. How could a puny human think he might survive there? Then our plane passed below the clouds, dropping us back into civilization and the myriad mundane details of launching an expedition. We had come to measure Mount Everest for *National Geographic* and, as all climbers want to do, measure ourselves. One step at a time, literally and figuratively, we had worked our way back up to what we had seen from the plane, and now were poised on the precarious triangular edge of that foreboding pyramid. We knew we didn't belong in this place; no one did. The view from the plane had told us that.

Everest is a giant among the innumerable jagged-edged mountains that intrude into the sky in this part of the world. The Himalayas are young and still growing. The process by which they formed began some 80 million years ago when pieces of the earth's single gigantic landmass began breaking up and floating out across the ocean. The massive piece, or continental plate, we now call India drifted north and, 30 million years later, collided with the even more massive Asian plate. The leading edge of India was driven under the southern edge of Asia, which began to rise and form 'wrinkles'. These wrinkles are now the Himalayan mountain range, stretching for 2,414 km, 805 km in width, and containing all the highest mountains in the world. India is still sliding under Asia, so the Himalayas are still rising, at the incredibly rapid rate (for geologists) of about one-half centimetre a year. We were here to measure this rate, and the dynamics that create it, as precisely as we could, as well as to map Everest and determine its exact height.

It takes tons of equipment to measure centimetres on Everest. Our expedition had added laser telescopes, global positioning satellite beacons and computers to the usual mix of ropes, tents and climbing gear. Twenty porters were hired and forty yaks were rented to carry our equipment – I needed four yaks for my medical supplies alone. We started at an altitude of 2,743 metres, on the landing strip in the foothills where our supply plane had deposited us, then proceeded on foot through some of the most remote, desolate and spectacularly beautiful terrain in the world. Narrow, precipitous trails led between ice-covered mountains that would blend into the clouds and then suddenly reappear above them against the brilliant blue sky. Because the land has been steadily uplifting for 50 million years, glacial runoff has carved out deep sheer gorges. These are spanned at dizzying heights by rickety wood-and-rope suspension bridges that sway in the wind as you cross them. The rotted-out floorboards and frayed ropes provide up-close evidence of the power of this hostile environment – and of the absence of any recent maintenance inspection.

We made our way up one side of a mountain, then down the other, and then up the next, gradually making net gains in altitude. Too rapid an advance up the mountain would bring on acute mountain sickness, the most common high-altitude problem encountered by lowlanders. The symptoms – a throbbing headache and nausea – are very similar to a hangover, and the cause is probably the same too: dilation of blood vessels and a shift of fluid into the brain that increases pressure within the skull. In a bar, vessels dilate in response to too much alcohol; on a mountain, it's in response to too little oxygen. Strong coffee helps the hangover because caffeine stretches blood vessels. For acute mountain sickness, the treatment is to go back down a way, or at least stop going up, until the vessels re-equilibrate, which usually takes a day or two. It can be prevented altogether by simply going up slowly, a lesson that skiers learn the hard way when they fly from sea level to a mountain resort and expect to ski the next morning. Some never learn it at all and are left with the mistaken impression that they can't tolerate altitude, when in fact all they have to do is go slower.

Our expedition stopped early each afternoon, and wherever we were, crowds of Sherpas formed outside my tent, some coming from more than a day's walk away to take advantage of the opportunity to see a Western doctor. Though none suffered from acute mountain sickness, there were plenty of other illnesses. The most common complaints were stomach pains, largely due to bad diet, and coughing. Sherpas don't have chimneys in their huts, so smoke accumulates inside, leading to chronic lung conditions such as emphysema. Women brought their entire families, not wanting to lose the chance to have each of their children examined. I diagnosed one child with pneumonia. Lacking paediatric medications, I crushed antibiotic tablets with a stone, then divided the powder into paediatric doses individually wrapped in toilet paper. There were also a lot of dental problems. One woman had a rotten tooth, which I prepared to extract by giving her a lidocaine injection. Once the tooth was numb, however, the pain went away and the woman believed she was cured. No amount of talking could convince her otherwise; she refused to let me pull the tooth.

I did what I could for everyone I saw and, in the last village we passed through, the people showed their gratitude by placing my stethoscope on a stone altar where it was blessed by the local lama.

We continued to march upward, beyond the villages and trails and on to the Khumbu Glacier, following it to its source – the slope of Mount Everest – where we set up base camp. At 5,334 metres, we were already higher than any point in the contiguous United States yet only at the foot of Everest, with still over 3,353 metres to go. We had done the easy part, climbing 2,743 metres in ten days, a fairly slow rate of ascent. Coming up any faster, however, would have killed us. Though we were all short of breath, tired, and cold we could now survive at this altitude almost indefinitely. Had we been dropped off here by helicopter, we would all have been dead within a few hours. The ten days had given our bodies the time to acclimatize – to make internal adjustments that compensate for the lack of oxygen.

The air in our atmosphere consists of 21 per cent oxygen, whether at the earth's surface or at the fringes of outer space. Because air has weight, it presses down on the air below it, creating pressure. The higher up the air is in the atmosphere, the less air there is above it, so the less air pressure. Air at 5,334 metres (base camp on Mount Everest) is under only half the pressure of air at sea level. At 8,839 metres (the summit of Mount Everest) the air pressure is only one-third that at sea level. Lungs depend on that pressure to force air inside.

Muscles expand lung walls to create a vacuum that's filled by air, which rushes in through the trachea and bronchial tubes to inflate the air sacs, or alveoli. When outside pressure drops too low, the air just meanders in and may not reach some of the alveoli at all.

Supplying oxygen to the body depends on getting it from the alveoli into the blood. A network of tiny blood vessels called capillaries surrounds each alveolus like a fine mesh. Both the capillary and the alveolus have very delicate membranes stuck to each other, forming a thin wall. Oxygen moves easily through this wall by a process known as diffusion. Diffusion occurs passively, its speed controlled solely by the difference in pressure between the two sides. Lower pressure in the lungs means a slower passage of oxygen into the blood. Arterial blood – fresh blood leaving the lungs – is normally almost fully saturated with oxygen. Two large arteries in the neck, known as the carotids, bring some of that blood directly to the brain, a voracious consumer of oxygen. Within those arteries are nerve sensors called carotid bodies, which monitor the oxygen content of the blood flowing by. Should the carotid bodies detect any decrease, they send a signal to the ever-vigilant hypothalamus, which reacts by inciting responses to increase the oxygen supply. Lungs expand more widely to create a greater vacuum, thereby re-establishing a more effective differential between themselves and the diminished outside air pressure. The heart pumps faster to absorb and distribute the available oxygen more rapidly.

Deeper breathing and a faster pulse were exactly what I felt at the start of my summit attempt after I shut off my oxygen tank and stepped out of my tent into the air on the South Col. The oxygen saturation in my blood suddenly dropped, probably to around 80 per cent. Later, when I ran out of oxygen on the Triangular Face, the air pressure was lower still and the drop in my oxygen saturation even more precipitous. I was too weak to stand, lost my peripheral vision and had the sensation of being detached from my body – the effects of oxygen loss on the muscles, eyes and brain. My saturation then was probably around 70 per cent. I was still capable of thinking, a function that cuts off at about 50 per cent. Should saturation fall below 30 to 40 per cent, unconsciousness and death ensue.

Other than by increasing breathing capacity and speeding up blood flow, the body has no immediate way to reverse an oxygen deficit and certainly no way that will enable it to survive if the drop is rapid or overwhelming, such as in an airplane that suddenly loses cabin pressure or after a helicopter ride from sea level to base camp. However, the body does have a highly effective long-term plan for dealing with gradually decreasing air pressure (up to a certain altitude limit), such as we experienced on our ten-day trek to Everest base camp (the base camp, coincidentally, lies at that limit). The components of the plan are what we call acclimatization.

The process begins a few hours after the hypothalamus realizes that the oxygen deficit is not being fully corrected by the heart and lungs and sends out signals for help to other organs. One of the first signals goes to, of all places, the kidneys. The kidneys act as the body's sewer system, filtering blood and controlling the discharge of waste products. One waste product over which it does not have control is carbon dioxide, the by-product of respiration, which is eliminated through the lungs. Deeper breathing washes greater amounts of carbon dioxide out of the blood, leaving more room to take in additional oxygen. But carbon dioxide is an acid, and if too much is lost, the blood becomes alkaline, disrupting its chemical reactions. The kidneys can't stop the release of carbon dioxide, but they can rebalance the blood chemistry by releasing more bicarbonate, an alkaline substance over which they have control. Urination increases until the necessary proportions are restored. The lungs can continue to work hard. Unfortunately, the additional water loss (not to mention the water vapour lost by the lungs) contributes to dehydration.

The kidneys have another critical role to play. They (as well as other organs, especially the lungs) release a hormone, erythropoietin, which travels to the bone marrow, the spongy tissue within the long bones of the arms and legs. Marrow is the factory for red blood cells, the oxygen carriers; erythropoietin stimulates their production. More red cells in the blood means more oxygen transport, but reaching full capacity takes weeks. It would be useful to have a 'quick injection system'. For some animals, that system is provided by the spleen, the body's reservoir of red cells. The spleens of deep-diving sea mammals can be contracted, releasing a reserve supply of red cells into the blood but, as with South Sea pearl divers who were once thought to have the same ability, no evidence currently exists that the phenomenon occurs in mountaineers.

Transporting more oxygen through the blood won't help if it can't reach the tissues. Red cells carry oxygen by binding it to an iron-bearing protein pigment called haemoglobin – it's this iron that makes the blood red. The attachment is made in the lungs when oxygen, under pressure, diffuses across the alveolus and is picked up by the passing red cell. Pressure is also needed to detach the oxygen at its destination so that it can diffuse out of the capillary and into the tissues. The transfer slows as the atmospheric pressure drops, but in response the body begins to manufacture more of an enzyme called 2,3, diphosphoglycerate (2,3,DPG), which weakens the oxygen-haemoglobin link and allows the oxygen to come off with less pressure. The higher the altitude, the more 2,3,DPG is produced, keeping pace with the decreasing pressure and preventing the whole oxygen transfer process from stalling. The tissues themselves do their part to help out: they grow more capillaries, making each cell closer to its oxygen source; they increase the size of their mitochondria, the tiny power plants within each cell in which the oxygen is actually burned; and they modify the enzymes controlling that burning – all to make the most efficient use possible of their decreased allotment of oxygen.

The body uses yet another trick to acclimatize itself, one it learned from the diving mammals. Muscles use a great deal of oxygen, and in emergencies they need a readily available supply. Muscles contain a protein called myoglobin, which is similar to haemoglobin but binds oxygen more loosely. Myoglobin provides an on-site oxygen supply, easily accessible for muscles that require extra energy in a hurry. Human muscles generally contain only small amounts of myoglobin; diving mammals stock it in large quantities and draw on the oxygen it contains while they are underwater. The storage of myoglobin in human muscle increases dramatically at high altitudes, providing some additional measure of protection against the tenuous air supply.

While our bodies were busy making all these adjustments to life at 5,334 metres, some of which take months or even years to complete, we were busy setting up base camp. Each of us had his or her own tent – on a long, stressful team expedition, a place for solitude becomes essential. There was a cook tent in which the Sherpas prepared our meals and a mess tent in which we ate them. No one could take a meal in his or her tent; we all had to come out and eat together – developing a feeling of camaraderie was as essential as the solitude.

As expedition doctor, my responsibility was to turn an empty tent into a medical facility and be prepared for every contingency. The tent was a domed structure, large enough to stand up in at the centre or for two or three people to lie side by side along the diameter. Foam mats on the floor provided insulation from the ice below for patients. To make shelves I stacked empty wooden oxygen crates along the walls. I hung carabiners from the roof support rods to serve as hooks to which I could attach IV bags. My four yakloads of supplies had been off-loaded and unceremoniously dumped on the tent floor.

I had given an enormous amount of thought to these supplies before leaving New York. Whatever I didn't bring I wouldn't have and would have no chance of getting. I had to be completely self-contained. Yet weight and bulk were limited; four yaks worth was about the most I could ask for. I had made a list of every injury and disease I might be called upon to treat, from head to toe, from fractured skull to athlete's foot, then written down every piece of equipment, supply and medication I would need to treat it – pulse oximeter to stethoscope, fibreglass cast to Band-Aid, cardiac stimulant to haemorrhoid pad. For all the medications, as well as for some of the instruments, I had to consider the effect of nightly freezing and daily thawing. It would be impossible to keep everything warm. Pills seemed safer than liquids and were lighter and less bulky. Each item had to be worth its weight. I brought surgical instruments but not a cardiac defibrillator. They are heavy and bulky, and any climber needing one was probably going to die anyway.

Even harder than deciding what to bring was deciding how much, particularly for heavy but critical items such as bags of IV fluid. A single climber in shock could easily go through ten bags of the stuff. How many climbers should I be prepared to save? I had to resign myself to the idea that disasters might occur that I would not be equipped to handle.

Once I had compiled my final list (ten typewritten pages), I solicited drug companies and surgical supply houses for 'free samples'. I was somewhat uncomfortable doing this since I generally try to ignore medical sales representatives. The conversations usually went something like this:

'I need a lot of pills.'

'How many? I've got ten or twenty in my case.'

'I need a thousand.'

When I explained that it was for an expedition to Mount Everest, that some would be used to treat local villagers en route and that the extra would be donated to a local clinic, the response was usually 'Wait a minute, I'll call my district manager.' And they always came through. Virtually all my medical supplies were donations. The entire collection was laid out in my office; my wife and children and I sat on the floor for days, repackaging and labelling each item.

Now I was sitting by myself on the floor of a tent, again taking inventory and checking each package for damage. I was especially anxious about our oxygen cylinders, which had been shipped directly from a factory in Russia; they are illegal in the

United States because they are made of only a thin layer of titanium – lightweight but prone to leaking and exploding. Most of the cylinders were earmarked for use high up on the mountain, but I needed to keep some at base camp with my medical supplies. Oxygen would be my drug of choice to treat most of the life-threatening illnesses in this environment.

I had divided all my supplies for transport so that if one of my four yaks wandered away or slipped off a narrow trail, my ability to deliver treatment wouldn't be severely compromised. Now I methodically regrouped everything according to category and set aside one complete set of every item I would need to treat every critical condition I might face. This took days. I worked slowly because I was getting out of breath easily and my head wasn't totally clear. Acclimatization was not yet complete. My body was becoming more efficient at using oxygen, but it would still be slowly deteriorating even when maximally acclimatized. That deterioration is the reason why nobody lives here.

Base camp is the physiological border for the human body. Going any higher now would be fatal. Once we were fully acclimatized, though, we would be able to make brief forays across the border to higher altitudes. Short exposures to even thinner air would serve as a stimulus to push each body defence to its maximum and, though further acclimatization would no longer be possible, they would increase the body's ability to withstand the deadly environment a little longer. Hence the physiological basis for the climber's intuitive tactic to 'climb high, sleep low' – reach maximum altitude during the day and return to a safer altitude at night. But Everest is too big a mountain for a day trip from base camp. To reach the top we needed outposts along the way: four camps to which we could retreat as we pushed toward the summit. Establishing each camp required bringing supplies up a little at a time from base camp. The expedition's need to make repeated trips to higher and higher altitudes in order to stock the camps coincides exactly with the body's need for brief exposure to progressively less and less oxygen.

After four days the medical tent was ready to function – and so were our bodies. We began moving up supplies to build the lower camps, staying overnight, then returning to base. Teams that had been there longer had already progressed much higher. A four-man Indian expedition, which had left base camp several days earlier, was moving up from Camp III at 7,010 metres to Camp IV at 7,925 metres, the camp before the summit. The section of the route they were on crosses a sheer ice face spanned by horizontal ropes anchored in the ice, to which climbers hook their harnesses as they traverse the slope. The attachment prevents a 914 metre slide, but any climber who slips will be left dangling from the end of a rope, completely exposed to the ice, wind and cold. Many teams use oxygen to make the traverse. The Indians, however, had a very limited supply, which they had already stashed at Camp IV for use on their summit attempt. To further conserve oxygen, they started their traverse late in the morning so that they would spend less time at Camp IV, where they would need to be breathing oxygen all night. Though they had dressed as warmly as they could, their clothes were barely adequate, and they carried no radios. They would soon learn that they had cut their safety margin too thin.

As the day wore on, the temperature fell. Freezing winds swept over the ice, carrying away the climbers' body heat. At first they felt cold, then painfully cold as their bodies sent progressively more urgent signals to their conscious minds to do something. But midway across the smooth, almost featureless slope to which they were closely attached by a short rope, the climbers had no chance to find shelter quickly. Their body temperatures dropped and they started shivering – aimless muscular bursts that begin in the trunk and arms for the sole purpose of generating heat. The contractions spread to the jaw muscles, and their teeth started chattering. Shivering produces only about as much heat as walking. Vigorous motion of the large muscles of the arms and legs constitutes a far more effective way of combating the cold, especially if the power generated leads the person out of danger.

When heat production can no longer fend off the cold, the body conserves its warmth by constricting blood vessels in the areas that leak the most heat. Hands and feet, noses and ears become pale and cold. The head and neck is another highly exposed area, but the body plays favourites. Despite the heat loss, flow to the head remains high in order to assure an adequate supply of blood to the brain. Scientists have only recently discovered the prioritizing of body parts; mothers, who insist their children wear hats and scarves on the coldest days, have long known it.

With nothing more to put on and no way to block the wind, the four Indian climbers tried to traverse faster along the rope. They were generating heat and moving toward shelter but also burning large amounts of fuel reserves – a process the body can't sustain for long. The only other protective response humans have against cold is goose bumps – which occur when the tiny muscles attached to hair follicles contract to straighten our body hairs, creating loft to trap warm air against the skin. It works for the feathers of high-flying birds and the fur of arctic foxes. For humans, with their relative paucity of hair, the mechanism seems almost pathetic.

Our bodies were designed for the tropics and are woefully inadequate to defend against cold. We survive only because of clothing and shelter, making us dependent on brainpower and manual agility. For every 0.6°C drop in body temperature, cerebral metabolism decreases by 5 per cent. Chemical reactions slow. Thinking becomes sluggish, and fine motor dexterity is lost, critically affecting our ability to get out of the very situation that is causing the problem. As we run out of energy to keep the internal fire burning, the cold takes over. Electrical transmission of nerve impulses is delayed, its amplitude reduced. Body parts become numb, limbs lose their coordination, the mind becomes apathetic.

Two of the Indian climbers stopped to rest on the ropes. The other two continued slowly upward. None of them thought to turn around. As their body temperatures dropped below 32°C, they lost even the energy to continue shivering. Once shivering stops, internal temperature drops precipitously and the body quickly spirals downward. Cold muscles lose their elasticity. Lungs can't expand and limbs stiffen. Muscle activity is reduced, decreasing heat generation still further. Soon the climbers would not be able to move at all.

One of the two climbers resting on the rope had enough presence of mind and enough coordination to take his sleeping bag out of his pack and place it upside down over his head and chest. Because his harness was attached at his waist to the rope in the ice, he couldn't get the sleeping bag any lower. The other climber just hung from the rope, exhausted. By about 3 a.m. the climber with the sleeping bag over his head had regained some strength. He was unable to see his teammate in the total darkness but managed to descend to Camp III. There he found a radio and called down to base camp.

When we heard the news, we realized that some members of our team were also at Camp III, having spent the night there after dropping off supplies. We radioed up to Todd Burleson, our expedition leader. He made his way over to the Indian tent and found the exhausted, hypothermic climber lying there. Todd brought him to our team's tent, and while he was being warmed and fed, the climber told Todd that just before reaching Camp III he had looked back up the slope and, in the dawn light, saw his teammate dangling from the rope. 'His arms were moving, not like from the wind.' He believed he might still be alive.

Before Todd and the other climbers at Camp III could mount a rescue attempt, however, we got information that made such an attempt unnecessary. One of the two Indian climbers who had continued upward descended back to camp this morning. He had reached Camp IV the previous evening. His partner had also reached Camp IV, staggering in some time later. His body temperature was probably above 29°C, for he was still conscious and able to move, but it was also clear that he had exceeded his physical limit. His body defences had been overwhelmed; he couldn't summon the energy to produce heat. Without some external heat source to rewarm him, death was inevitable. His temperature continued to drop. When it fell below 29°C he became unconscious, thereby losing any last chance to do something to save himself. His heart was still beating regularly, though only once or twice a minute. Below 27°C the feeble heartbeat became an irregular flutter, and at around 21°C it stopped altogether.

In the morning the surviving climber left his lifeless partner in the tent and came back down the ropes. He passed the body of the other exhausted climber, still resting in his harness, frozen to death.

'He was our expedition leader,' the climber told us. 'His dream was to climb Everest.'

A dream may spring from the mind, but it needs a brain to contain it and a living body to sustain it. For the expedition leader, the dream was extinguished when the chemical sequences stopped and the fires went out. As for the rest of us on the mountain, our bodies were still more or less protecting us from our surroundings, incubating all the chemical reactions that create heat, heartbeat, motion, emotion, ideas and dreams.

It's uniquely human that the most subtle of those reactions – the ones that produce (or perhaps form as the result of) our abstract thoughts – are able to override our most powerful instincts for self-preservation. A human body can be motivated to take risks not just for practical necessities, such as finding food or escaping predators, but on behalf of abstract concepts like scientific curiosity, adventure and individual achievement, which have nothing to do with survival. When the most highly developed part of the brain, the frontal lobes – the seats of judgement, thought and will – give a command to enter a hostile environment, the rest of the body suppresses its survival instinct and dutifully carries out the command, protecting itself from the results as best it can.

Human bodies vary enormously in their ability to withstand cold. The Yaga Indians of Tierra del Fuego don't wear clothes, even in winter, and the women have been known to calmly breast-feed their infants in open boats during ice storms. Aborigines in Australia and Kalahari bushmen in Africa are able to sleep naked on the ground during freezing desert nights. These desert dwellers have developed the same tactic as camels, though applied in reverse: they allow their body temperature to move several degrees closer to the outside temperature, making body temperature that much easier to maintain, since the differential has been decreased. Camels save energy by not sweating, allowing their temperature to rise during the day. They retain the excess heat until evening, when the desert breezes cool them off again. For the human desert dwellers, who can't store heat the way camels do, the same cold night breezes pose the threat of hypothermia. They save energy by not shivering, lowering their metabolism and allowing their temperature to fall overnight to as low as 35°C. It's almost a mini-hibernation. These people awaken with no adverse effects and let the morning do the work of rewarming them.

The rest of us can't get away with this. We would become hypothermic and perhaps not get up at all. How some people do it is something of a mystery. It may well involve the same chaperone proteins that increase heat resistance in marathon runners. The formation of these proteins is stimulated by cold as well as by heat, and once formed they will prevent other proteins from being deformed, or 'unfolded', and inactivated by cold. Such tolerance does not appear to be due to any inbred gene; rather, it stems from maximizing the body's ability to adapt to cold through repeated exposures. Before his Antarctic traverse, French explorer Jean-Louis Etienne would begin shivering as soon as his body temperature started to drop, just like any other Frenchman. On his return, however, he discovered that he could tolerate a body-temperature drop to 35°C before shivering commenced. Though the exact mechanism remains mysterious, the idea that the body can 'learn' to tolerate cold has been an extreme medicine assumption for a long time. In preparation for their voyages, early Arctic explorers were advised to take cold showers – increasing their duration and decreasing the temperature every day until departure. Further advice was to stop washing as soon as they had left. Dirt accumulating on skin helps the body retain heat. This advice has also proven to be true, and I know from experience that it's one rule mountaineers have no trouble following.

Eskimos are the premier cold weather adapters. Besides having their share of those chaperone proteins, they have a thick layer of subcutaneous fat that provides extra insulation. This seems to be a response to the environment rather than an inbred characteristic. People in temperate climates often take on weight during the winter and lose it during the summer. Eskimos tend to retain it. With animals being virtually the only food available, an Eskimo's diet is rich in animal fats, which, when metabolized, not only provide a great source of heat but can be easily converted to human fat and deposited below the skin. Living in such a harsh environment with such limited resources, Eskimos are provided with the only fuel efficient enough to allow them to survive there.

Eskimos also benefit from their shape. They have a compact design with short arms, legs, fingers and toes. They're styled for the Arctic much the same way penguins are styled for the Antarctic. However, humans can't compete with animals in the extent of their adaptations. Our muscles stop contracting below about 21°C for example, whereas the muscles of the arctic fox continue to work down to nearly –18°C. Caribou have fat whose composition varies with its location. Within the body proper, the fats are saturated – like butter, able to remain liquid only when the temperature is high. As the fats become more exposed, such as they are in the legs, they become progressively more unsaturated – like olive oil, which doesn't freeze even at low temperatures. No animal can beat the wood frog, though. When the temperature drops below zero, the animal freezes. Then, when the temperature rises again, it simply thaws out and goes on about its business.

Evolution has left humans well short of matching those talents. We developed big brains instead – either because we needed one to make up for our limited physical abilities or because, once we had one, we learned to get by with what we had. Either way, humans rely primarily on behaviour to survive the cold, mimicking animals and often outperforming them, using what's available in the environment. Sherpas build stone huts and often sleep in people piles to stay warm. Eskimos build igloos and wear animal skins inside out – because wearing the fur that way traps more warm air than the way the animals do it. The ultimate human behaviour modification that helps us survive cold is migration. Humans retreated to equatorial zones during the last Ice Age. Many still head south every winter to escape the cold.

That last thought seemed especially relevant to those of us who often wondered what we were doing on Mount Everest. Life at base camp was like living in a refrigerator. We had ice below and on all sides of us. The supercooled, superdry air circulated over the ice and then through our lungs. Each breath had to be warmed to body temperature and moisturized to 100 per cent humidity to keep the lungs happy. The heat and water required to do this were far more than our bodies could provide, so our air passages, the trachea and bronchi, became irritated and dry, giving rise to what climbers call the 'Khumbu cough'. Everyone who climbs Everest gets it. Though I handed out throat lozenges and cough syrup and told everyone to breathe over a pot of steaming water, the only real cure is descending. Had I sent down everyone who had a cough, there would have been no one left to climb. But had I not sent anyone down, at least temporarily, some would have been debilitated from the repetitive spasms, a few would have broken ribs from the forceful exertion and one or two would have died from pulmonary oedema.

It was no easy job to sort through the chorus of coughers who collected for dinner each evening, making the mess tent sound like a tuberculosis ward. On one particularly noisy night, after I had listened with my stethoscope to all the loudest coughers, I handed out a lot of medication. Everyone seemed to need it. Even our kitchen boy, Koncha, who had been working hard around our table all night, asked for pills. I hadn't heard him cough even once but gave him some pills anyway, thinking that he just wanted some attention.

Just before sunrise, I was awakened by an urgent shout just outside my tent.

'Dr Ken! Come quick, come now!'

Outside, I found our cook, Ong-Chu, pointing to the kitchen tent with one hand, pointing to his chest with the other and shaking his head. I entered the kitchen tent and, in the yellow light of the lantern, saw Koncha. He was in his usual sleeping spot, a mat in the far corner of the tent, but was sitting upright, making great heaving efforts to breathe. Even from across the tent I could hear wet bubbling sounds coming from his chest. His breathing was shallow and rapid, interrupted by violent bursts of coughing that brought up blood-tinged globs of frothy mucus. His lips were puffy and blue, his skin was clammy and pale and his eyes were wide and terrified. This was pulmonary oedema – a high-altitude killer. In the early days of extreme medicine, it was thought to be caused by the breath of dragons lurking behind high mountain passes. Since the discovery of oxygen, however, this theory has been largely disproven, although we're still not sure exactly what causes it. Pulmonary oedema only occurs at altitudes above 2,438 metres, and while it comes on suddenly, it does so only after its victim has been there at least a few days. Usually it develops from a cough or after heavy exertion, and rapidly worsens at night or in bad weather, making treatment and evacuation to a lower altitude all the more problematic.

Healthy lungs have it easy at low altitudes, where there is more than enough air pressure to fill the alveoli and drive oxygen into the blood. Lungs take in far more oxygen than they consume, and thus provide a large surplus for every other organ in the body. They even have room to spare: there is so much excess capacity that many alveoli remain closed during normal respiration. As altitude increases, however, the consequent drop in pressure makes the lungs work harder. When the blood and then all the other organs signal the hypothalamus that they aren't getting enough oxygen, the hypothalamus forwards the complaints to the lungs. As altitude increases further, the complaints get louder and the lungs become the most put-upon organ in the body. Their first reaction is to breathe faster and deeper. This out-of-breath response speeds up air intake and opens dormant alveoli. To benefit from the increased airflow, the lungs must keep pace by increasing the blood flow to match it. This they do by turning up the pressure in the pulmonary artery, the conduit that carries blood from the heart to the lungs. To make oxygen intake more efficient, the lungs also have another mechanism, one that works well at sea level but backfires at high altitude. Blood flow through each section of the lung is monitored for oxygen content. Where alveoli are not providing enough oxygen, blood vessels constrict so that flow can be diverted to more 'productive' areas. This feedback system evolved to help humans get by when injury or infection damaged a section of lung. Those are threats the body understands. It doesn't understand low oxygen pressure, however. Lungs were never designed to function at high altitude.

Faced with low oxygen readings everywhere, each section of the lung reacts as if it were the only one affected, narrowing its vessels to divert flow to some other section. But the vessels are constricted everywhere – there's nowhere for the blood to go. To make matters worse, the pulmonary artery is delivering blood at high pressure, still trying to keep pace with the high rate of airflow. The effect is like opening the valve to a garden hose and then pinching the end of the hose shut. The flow comes through the vessels as a narrow jet, coursing into the delicate alveolar capillaries under tremendous pressure. The thin membrane wall separating the capillary from the alveolus begins to leak fluid and then ruptures like a bursting dam. Plasma, the liquid part of the blood, spills into the air sacs. The victim drowns in his own fluids.

This was precisely what Koncha was about to do unless I could stop him. He needed to be moved to the medical tent. Two Sherpas began to help him up, but I insisted he be carried, though the tent was only a short distance away over the ice. The exertion of walking would increase his pulmonary artery pressure even more.

The sun was still not up. The medical tent was dim except for one powerful flashlight. Koncha was deposited on two foam mats propped against an oxygen crate, centre stage beneath the spotlight. Kneeling down next to him, amid supplies and an audience of anxious volunteers, I took out a stethoscope, although I hardly needed one to hear the turbulence within his chest – the gurgling noise of air bubbling through fluid filled three-quarters of his lungs. Struggling to maintain blood and air flow, his pulse had doubled and his respiration rate tripled. The oxygen saturation in his blood (normally close to 100 per cent) had fallen to 26 per cent – a level that until then I had thought incompatible with life. Not many of his alveoli were still working. We kept him sitting upright so that the fluid pouring into his lungs would fall to the bottom, hoping that would leave the topmost alveoli high and dry long enough for the leakage to stop and the water level to recede.

The treatment is to reverse the cause – to reduce the force squeezing fluid through the capillaries into the alveoli. I placed a mask over Koncha's face so that the air he inhaled, though still at atmospheric pressure, was 100 per cent oxygen instead of the normal 21 per cent. Body sensors immediately detected the fivefold increase in concentration and eased off on their distress signals to the lungs and heart. Pulse and respiration rate dropped. The pulmonary artery pressure valve relaxed. There was more oxygen around now. Nonetheless, many of the alveoli were underwater, not able to let the oxygen through. Responding to the low oxygen level, the vessels surrounding them remained constricted, so the fluid pressure against their walls remained high and they were still leaking. Koncha's breathing remained laboured, and he was still coughing up blood and bubbles. I had wanted to start an IV immediately, but that early in the morning the bags of solution were frozen solid, so I had sent a few to the kitchen for defrosting. The first ones came back. I hadn't been wearing gloves, so my hands were too cold to judge the temperature of the bags, but by touching them to my cheek I verified that I wasn't going to fill Koncha with ice water.

Koncha was trembling so violently that two people had to hold him down while I inserted a catheter in his vein, connected the IV tubing and started giving him fluids. His lungs may have been flooded, but the rest of his body was dehydrated – in part from supplying all that fluid to his lungs. Having an IV in place made it much safer to use nifedipine, a drug that relaxes the muscles of small blood vessels. It would help relieve the constrictions that had formed in Koncha's lungs in response to his flooded alveoli but it would also dilate vessels throughout the body. The total effect of so much lost resistance sometimes means a precipitous drop in blood pressure. In that case I would now be able to keep Koncha pumped up by adding more fluid through the IV.

With oxygen driving down the pulmonary artery pressure and nifedipine opening the vascular channels, the blood flow was slowed. The pinch was eased. Koncha's lungs stopped leaking. His coughing subsided, the blue colour left his lips and the fear left his eyes. He was on high maintenance, though. Without the oxygen, fluids and medication, his symptoms would come back. We had to get him out of there.

Mount Everest's base camp is at the altitude limit for a helicopter evacuation, and for it to be possible, conditions have to be perfect – which, on that day, they weren't. The sun was up, but the sky was overcast and it was snowing lightly. We needed to get Koncha lower – even a few hundred metres would probably provide enough air pressure to keep his lungs inflated. The plan was to carry him off the glacier, along the trails that pass through the villages. There was a small clinic located at 4,267 metres; if he wasn't better before then we would go there.

Koncha was breathing easier now and alert, but if we let him walk, the exertion would drive his pulmonary artery pressure back up so rapidly he would collapse before he got out of camp. He would have to be carried, but until we were sure his lungs were dry, he had to remain sitting up. I was pondering how to do this, but the answer was immediately obvious to the Sherpas. A lightweight aluminium folding chair was opened and placed face-out on the back of a volunteer. Ropes were passed around the frame and tied to his waist and shoulders. Another rope was looped from the chair top across his forehead. The Sherpa squatted down so that the chair legs touched the ground, and Koncha sat down. Once Koncha was strapped in, the Sherpa stood up, absorbing the strain fairly easily as the ropes tightened around his abdomen, chest and forehead. Another Sherpa loaded an oxygen cylinder into his backpack, winding its hose forward around the front so that the regulator could be placed over Koncha's mouth. He was using the longest hose we had, but staying tethered to Koncha would not be easy. Two other Sherpas loaded extra oxygen tanks along with some of my medical supplies. Our mountain ambulance was ready.

We travelled much faster than I expected. Taking turns carrying Koncha, the four Sherpas set a pace I couldn't match. I routinely fell behind, catching up only when they stopped to exchange loads. In a few hours we were off the glacier and on to the trails. Koncha was holding his own but still dependent on oxygen. Though we had unhooked his IV for the trip, I gave him some more nifedipine. He looked as if he could use it. We hadn't yet lost much altitude. The next part of the trail down was steeper; I hoped he would begin to improve more rapidly.

After several hours, we began to see signs of civilization – a yak, a few more yaks, a yak herder, then finally a village – Pheriche, which, at 4,267 metres, is the highest permanently occupied settlement in the Himalayas: a few stone huts and two teahouses to accommodate visiting hikers. We were greeted incredulously as we came into town by a group of young female German trekkers, resting in front of one of the teahouses. Koncha was breathing much easier now, taking only occasional puffs of oxygen and enjoying the attention he was receiving from the girls. The warm teahouse seemed like a good place to spend the night. The rooms were all taken, but there was space available on the floor. Two of the German girls made sure Koncha was fed, then arranged some mats on the floor for him to sleep on. I appreciated the unexpected nursing help and was pleased at Koncha's improvement. Pulmonary oedema often worsens at night, however, and I was anxious to see how he would be in the morning.

The first thing I did when I woke up was look over to make sure Koncha was still alive. His mat was empty. I found him in the kitchen, sipping tea and flirting with the trekkers. He didn't speak German and they didn't speak Nepali, but it seemed to me he was asking one of the girls for a date. Clearly, Koncha needed no further treatment. The 1,067 metre drop had brought him to an altitude where air pressure is high enough to sustain life. As long as he remained below this invisible barrier, he would be safe. If he dared to cross it again he would be courting disaster.

Koncha had been at base camp for weeks before suddenly developing pulmonary oedema. True to form, it had struck at night, when the barometric pressure routinely drops; and earlier that evening he had been exerting himself around the kitchen. Bad weather had been rolling in – a front of low pressure. What that meant, effectively, was that the altitude of base camp rose. The additional drop in pressure pushed Koncha, already at the limit of his physiology, over the edge. But he had endured higher altitudes and stormier weather many times before. What happened this time? Maybe it really was dragon breath.

Leaving Koncha with his new-found friends, I re-entered the realm of pulmonary oedema, the no-man's land above 4,267 metres, once traversed by Mongol warriors but now frequented more by climbers and trekkers. Over seven hundred years ago, Tibetan doctors described symptoms of 'oppression of the chest' and 'white froth coming from the mouth' among the Mongol invaders. They also noted, though they couldn't explain why, that the malady seemed to attack only strangers. Modern doctors have confirmed the observations of their early Tibetan counterparts, so I was puzzled why Koncha was struck rather than a visitor like me. I should have been much more vulnerable. I made my way back up to base camp pondering this question and watching for dragons.

When I arrived I mentioned to our cook, Ong-Chu, that I had never before known a Sherpa to develop pulmonary oedema.

'Koncha is not a Sherpa,' Ong-Chu informed me. 'He's a Rai. He has been across high mountain passes many times, and he wanted a chance to make big money as a kitchen boy, so I gave him the job.'

Ong-Chu had solved the riddle. Rais are an ethnic group from the lowlands of Nepal. The Sherpas are of Tibetan stock, having migrated from the high plateaus into the Himalayas about five hundred years ago to escape those rampaging Mongols. Their ability to withstand the altitude and cold is the result of natural selection through uncountable generations, combined with maximum adaptation by each individual acquired from a lifetime in the mountains. Exactly how they do it is still largely a mystery. Some possibilities seem obvious: larger lung capacity to augment oxygen intake, a bigger heart to circulate blood faster, a lot more red blood cells to increase oxygen-carrying capacity. But obvious answers are often wrong. The Sherpas have undergone none of those design changes. Their adjustments to high altitude are far subtler. Nonetheless, taken together, they produce a body exquisitely fine-tuned for life at the upper edge.

Sherpa lungs are not especially big or powerful, but they are very sensitive to low oxygen and will maintain an increased rate of breathing even at rest – that is, they have a higher idling speed. Sherpa hearts are not stronger, but they don't have to be. To keep pumping, the heart muscle normally relies on fat, a fuel that doesn't burn well in low oxygen. A Sherpa's heart muscle can take in glucose, which it can burn 50 per cent more efficiently, thereby providing the same cardiac output with much less work. A Sherpa's blood has only a few additional red cells, but its capacity to transport oxygen is greatly increased by special enzymes that ride along with the haemoglobin to speed up the capture and release of oxygen.

Sherpas have other advantages as well. Their lungs are capable of producing large quantities of nitric oxide, a naturally occurring chemical in all human lungs, but one Sherpas produce twice as much of as almost everyone else. Nitric oxide acts as a vasodilator, opening up constricted vessels in the lungs. Having a large supply might be why Sherpas at high altitudes are nearly immune to pulmonary oedema.

One survival mechanism all humans share involves a preprogrammed sequence of chemical reactions called anaerobic respiration. Though it is not as efficient as normal respiration and results in the buildup of lactic acid, which has to be broken down later, anaerobic respiration provides quick energy for emergencies without using oxygen. It would conceivably provide a made-to-order system that Sherpas could exploit to adapt to their environment, and a possible explanation for why they are able to work so hard at such high altitudes. A neat theory but a wrong one. Sherpas working at maximum capacity produce less lactic acid than the rest of us, a condition known as the lactic acid paradox. The probable explanation is that Sherpas don't need to form lactic acid because they can maintain high fuel-burning efficiency even in low oxygen – by using enzymes we still haven't identified.

When species are subdivided and isolated from one another, adaptive responses to a similar extreme environment may evolve very differently. The Sherpas of the Himalayas and the Quechua and Ayamara Indians of the Andes last had a common ancestor about two hundred and fifty thousand years ago. People have been in the plateaus of Asia for at least thirty thousand years, but the Quechua and Ayamara took a while to migrate to the mountains of South America, arriving in the Andes only about ten thousand years ago. Since then, they've been faced with the same environmental challenges as Sherpas, but their bodies have responded in nearly opposite ways.

The Andes natives are barrel-chested and have large lungs with extra alveoli and more capillaries. Their hearts are bigger, with well-developed muscles in the right ventricle – the chamber that pumps blood under high pressure through the pulmonary artery into the lungs. They have increased numbers of red blood cells to transport oxygen and an increased volume of plasma in which to float them so the blood doesn't sludge. Their adaptations seem to be acclimatization pushed to the limits – mechanical changes rather than the biochemical ones of the Sherpas. They do, however, show some adaptations at the molecular level. Like the Sherpas, they demonstrate the lactic acid paradox, and their hearts utilize glucose. Their lungs produce extra nitric oxide, though not as much as the Sherpas do. However, they still do not share the Sherpas' increased lung sensitivity to low oxygen or their special haemoglobin-binding enzymes.

But both systems – mechanical and biochemical – work, and not just for humans. There are other large mammals beautifully adapted to life in the high mountains. Sherpas share their extreme environment with yaks; the Quechua and Ayamara Indians, with llamas. Yaks have developed haemoglobin-binding enzymes similar to those of the Sherpas; llamas have strong right ventricle muscles, just as the Indians do. They don't seem to possess the other human modifications, but they do fine as long as they remain at altitudes that are within their physiological borders. Only humans push themselves beyond their bodies' limits for reasons that have nothing to do with survival.

On Mount Aucanquilcha in the Andes, there is a permanent sulphur mining settlement at 5,334 metres – about the same altitude as Everest base camp. The mine itself is at 5,944 metres, but the miners find life there too uncomfortable to stay continuously. They prefer to sleep at the lower elevation, though that means a daily 610-metre climb. Sherpas are also unable to stay above base camp for any protracted length of time, and need to come down to recuperate after each short stay at a higher camp. Indians and Sherpas have taken different routes to high-altitude adaptation, but both seem to have arrived at the same physiological limit.

Although there are more permanent settlements at higher elevations in the Andes than in the Himalayas, the Sherpas are generally considered to be more smoothly adapted to their environment. They've had a 20,000-year head start. If whatever variations exist between the Sherpas and the Andes natives are the result of different gene pools, the two groups' high-altitude adaptability will continue to develop along different lines. If these adaptations are the long-term result of environmental pressure shaping natural selection and individual development, then perhaps the Indians will evolve in the direction of the Sherpas. However, the decreasing isolation and increasing mobility of both these groups make it far more likely that homogenization of genes and reduced environmental exposure will prevent either outcome.

Now that I knew that Koncha wasn't a Sherpa, or an Indian, or for that matter a yak or a llama, I understood how he could have developed pulmonary oedema, though I still didn't understand why. Other illnesses I faced were far easier to understand – and to treat. Besides coughs, headaches and diarrhoea, two other conditions unique to this environment afflicted my expedition. One was a high-altitude toothache. Air often gets trapped inside rotten teeth that have fillings. As the outside air pressure drops, the air in the tooth expands painfully. The treatment involves letting it out by inserting a hollow needle into the abscessed tooth. The other condition is high-altitude flatus. The cause is similar – air trapped in the many loops of the intestines expanding and forcing its way out the rectum – but the treatment is different: Pepto-Bismol and temporary isolation.

The conditions I have seen on Mount Everest have run from the mundane to the bizarre. I was about to enter my tent one afternoon when in the distance I spotted one of our Sherpa porters moving slowly and unsteadily up the glacier toward me. Heavy snow was blowing sideways against my face; it was too cold to wait outside. When the porter came into my tent he was wearing rubber thongs on his feet. He told me he had fallen through the ice very early that morning, sinking up to his knees in water. He'd had no way to dry himself off, and in fact had fallen through several more times before he reached solid ice. His feet were numb and swollen and he had lost control of his toes. 'All day,' he told me, 'I have been walking in someone else's feet.'

The porter's feet weren't frozen. In fact I was astounded to discover they were warmer to the touch than my fingertips. And they weren't white or blue; they were pink. He had been out in a snowstorm with wet feet, in subzero temperatures, for probably ten hours. Yet his feet were warm. I realized that the Sherpa was benefiting from a primitive yet carefully modulated limb-salvage reflex, called the 'hunter reflex,' present in warm-blooded animals but well developed only in those whose limbs habitually contact cold water, like wading birds in the Arctic and fishermen in Nova Scotia. The body's initial response to cold-water exposure of hands or feet is to constrict their blood flow in an effort to prevent rapid heat loss. If the animal or human can't, or won't, remove the body part, and it continues to cool, the blood vessels will dilate again, as if the body had decided that, on balance, losing a little extra heat was better than losing a hand or foot. With further cooling, the vessels begin to alternate periods of dilation and constriction, providing an intermittent blood supply as a compromise between trying to conserve both heat and limb. If the low-temperature conditions persist and the hypothalamus senses that body temperature is at risk, it will signal the vessels to remain constricted, effectively giving up on the limb and allowing it to freeze.

The feet before me were in the dilation-constriction phase. I needed only to dry them. Once they were out of the cold, the swelling subsided, the toes started moving, and sensation returned. I watched this demonstration of the body's ability to fend off the cold and reverse the damage, feeling that there was nothing much more a mere mortal could or should do. I offered the man a pair of my wool socks, but his feet had become so sensitive that he was unable to put them on. When the storm abated, he put his thongs back on over his bare feet and departed. He kept my socks, promising to use them next time. For me, wool socks were a necessity. For him, they were a luxury. He had been born and raised in this environment, and his feet, like the rest of his body, were highly resistant to cold. There is no Sherpa gene for cold tolerance (at least none has so far been isolated), and yet the porter was able to walk nearly barefoot over fresh snow because his body had had a lifetime to realize its innate potential to adapt. Any of his ancestors who hadn't possessed that ability would have been eliminated by natural selection long before the invention of wool socks.

With his feet self-restored, the porter moved down the glacier. Though he was far better adapted to this environment than I, he had no desire to go higher up into it. He was on one of the 'big headache' mountains that Sherpas know to stay away from, even if their employers will not. Gods live at the tops of these mountains and they draw out the strength and the breath of animals and humans who approach them. Those still strong enough to enter their higher reaches are warned away with terrible headaches. And if those warnings are ignored, the gods lose their patience. Punishment is swift.

Climbing to 6,401 metres a few days later, tired and breathless, I understood that the gods were annoyed at me. As I approached Camp II, I was hoping they wouldn't get any angrier. Before I had even reached my tent, I found out from climbers on my team who had arrived at the camp earlier that the gods had already inflicted their wrath on someone else. Rather than being greeted with the usual comments, such as

'Glad to see you' and 'Nice job', what I heard was, 'Man, are you slow. Go take a look at Ricardo. He looks worse than you.'

Ricardo was a South American climber with another expedition whom I had met years earlier on a mountain in Patagonia. His teammates told me this was his second trip hauling supplies up to Camp II. He had arrived yesterday, feeling fine and planning to move on up to Camp III this morning. But he awoke complaining of a pounding headache. He was given two codeine pills, and when that didn't work, he was given two more. His headache had eased, but he didn't feel well enough to come out of his tent.

Still wearing my climbing harness, I crawled in to see him. He was curled up in his sleeping bag, lying sideways. I expected he would remember me, and I knew he spoke English well, so I greeted him with a cheery 'Hi, Ricardo, how've you been?'

'What do you want?'

His hostility was a bad sign.

'Ricardo, who won the World Cup?' I asked.

'I did,' he replied.

Pressure was building inside Ricardo's brain. Memory and reasoning were impaired, as was control over primitive instincts like aggression. These functions are contained within the cerebral cortex – the outer, convoluted layer of the brain that fills the top, front and sides of the skull. The cerebral cortex was the last brain part to evolve and is responsible for all the higher thought processes. Extremely sensitive to oxygen deprivation, it's the first part to suffer when the brain swells enough to impede its own blood flow. When it malfunctions, people behave like Ricardo.

Cerebral oedema, swelling of the brain, is the worst punishment the mountain gods can inflict. As humans approach the gods' abode, organs react to the steadily dropping air pressure by dilating the vessels that feed them. This is both to encourage more blood to enter and to slow the flow so that there will be more time for the oxygen to be absorbed. The brain, being the body's priority organ, becomes the most engorged. It's also the only organ enclosed in a hard case.

Like a suit of armour, the skull protects against blows from the outside, but like a fortress, it blocks escape when the attack comes from within. As vessels fill with blood, an incompressible liquid, they push against the soft brain tissue, expanding it outward. The additional blood doesn't contain enough oxygen to satisfy the brain, so the swelling continues. The cerebrospinal fluid, a liquid bumper between the brain and skull that acts as an impact attenuator, absorbs some expansion. The fluid displaced by the swollen brain flows down into the spinal column, but soon there is no more room to expand, and the cerebral cortex gets pushed against the rigid, unyielding skull. As pressure builds, brain tissue is compressed, vessels are squeezed and fluid begins to leak out into the surrounding space. Besides adding to the volume and pressure, the fluid creates a barrier between the vessels and the brain cells, depriving them of what little oxygen remains. Cell membranes break down and fluid enters, swelling the individual cells and adding still more volume and pressure within the brain case. At sea level, vessel dilation is a survival adaptation. When man ventures too high, it becomes the enemy within.

Besides attacking the outer cortex and disrupting thought, cerebral oedema attacks the cerebellum, the part of the brain at the back of the skull that controls motor function. Fluid leakage and pressure buildup in the cerebellum cause a loss of balance and coordinated movement. At the base of the skull is the medulla, the most primitive part of the brain, containing the limbic system and the hypothalamus. The medulla connects to the spinal cord, processing raw sensory input and maintaining body functions. When pressure reaches down to this level, hallucinations begin and the body's most basic systems become deregulated. If this process is not rapidly reversed, death is certain.

We had to get Ricardo down.

'Get out of this tent,' I commanded him.

He fumbled with the zipper of his sleeping bag, unable to open it. 'I'm not leaving without my girlfriend,' he said.

Ricardo's girlfriend was in South America. We had to get him down now.

Far too much time had been lost treating Ricardo's 'headache'. Descent is the only truly effective treatment for cerebral oedema. Getting him to base camp would mean travelling through the treacherous icefall in the dark. Spending the night here, however, would be fatal. While Ricardo's teammates converted a ladder into a litter, my teammates brought me the fishing-tackle box filled with emergency medical supplies that I kept at Camp II. I gave Ricardo an injection of dexamethasone, a powerful steroid that reduces inflammation of tissues. It would shrink his brain for a few hours, temporarily relieving some of the pressure and, I hoped, improving his condition. His rescuers could inject him again en route to tide him over the rest of the way. The expedition had its own doctor at base camp; he could take the relay once they arrived.

We had no oxygen stored at Camp II, but we had the next best thing – a Gamoff bag, a portable hyperbaric chamber that weighs 5 kg. When folded up, the canvas bag can be carried like a backpack. When inflated by a foot pedal, it opens to become a body-sized cylinder. We put Ricardo inside and, with constant pumping, raised the air pressure around him so that his body would think it was 610 metres lower than it really was. We kept up the treatment until his rescuers were ready to leave. There was nothing more we could do for him. With rapid descent, the cause is removed and the symptoms usually ease.

Our expedition continued up to Camp III the following morning to position some supplies. It was three days before we got back to base camp. Before I could even ask, I was told that Ricardo had died. I went directly to the South American camp and found their doctor in the mess tent.

'He didn't die of cerebral oedema, you know.' The doctor poured me some tea and explained. 'He didn't look too bad when he got here, and I thought he would make it, but the next day he had a heart attack. Before he went up, I measured his haematocrit. It was seventy. I warned him not to climb, but he didn't listen to me.'

What the doctor was telling me was that Ricardo's blood contained nearly twice the normal concentration of red blood cells. High altitude had stimulated their production while blunting his sensation of thirst. Hard work in dry cold air had caused him to lose large amounts of water through exhaled vapour. With more cells and less fluid in which to dilute them, his blood was two-thirds solid, and flowing like tar. Fresh blood leaves the heart through the aorta, an aqueduct that divides, subdivides and branches off to every organ in the body. Right as the blood starts out, some gets siphoned off from the aorta and diverted into the coronary arteries – three vessels that return blood to feed the heart muscles themselves. The vessels are narrow and are too often made dangerously narrower when tightened by stress or clogged by cholesterol. Heart attacks occur when blood flow through these vessels is suddenly and totally blocked, usually due to severe spasm or deposition of an errant plaque of fat. At altitudes like this, however, where extreme medicine is the rule, heart attacks happen to young, healthy climbers because their blood has literally become too thick to fit through their arteries.

The cause of death on a high mountain can be as subtle as an overabundance of red blood cells or as brutal as a 610 metre fall. My team and I were methodically climbing a sheer ice slope below Camp III, perhaps two days away from Everest's summit. Kami, one of our support Sherpas, who had moved on ahead of us to drop supplies, had already reached the camp and was just starting his descent. Above me, I heard a sudden rumbling noise but my view was obscured by overhanging ice. For a long second, the noise grew louder and then a body flew over my head. We watched in horror as Kami bounced off the ice below us, tumbling and skidding down the slick slope before being catapulted over a crevasse and then landing, motionless, in the snow.

Kami had made a simple mistake. To save time on the descent, he had decided not to hook his harness into the safety line. Then he slipped, and his error in judgement cost him his life. Brains chronically low on oxygen function slowly. The result is fuzzy thinking. It can be overcome by intense concentration, but without enough oxygen, the extra brain power is hard to muster. Poor judgement comes naturally at high altitudes.

Deeply shaken, we pulled back to base camp to regroup and to do some painful soul searching. No measurements are worth a human life, but we had all assessed and accepted the risks and rewards of adventure. Stopping now wouldn't bring Kami back, and there was still valuable work to be done. Deploying the equipment *National Geographic* had given us would be an important step toward understanding the geology of the Asian continental plate. We decided to resume the expedition, pledging to be more careful than we had ever been before.

The reascent of the mountain took five days. We worked our way through the icefall to Camp I, then over the crevasses to Camp II. We reclimbed the ice slope to Camp III, passing the spot where Kami fell, then traversed to Camp IV at the South Col, the last camp before the top.

As we set out toward the summit – and toward that critical moment of decision – we had high hopes and willing muscles, but our strength and even our thoughts were dependent on the fuel in our bodies and the oxygen on our backs. Because the deep snow had slowed our progress throughout the night and into the morning, we had begun to run low on both. At an altitude of 8,565 metres, we had to make that critical decision whether to continue upward or to turn around. Poised on the precipitous edge of the southeast ridge, our lives quite possibly were hanging in the balance. The summit, 274 vertical metres above us, was tantalizingly close. Nonetheless, at our rate of progress, reaching it would probably take another five hours. And while we had enough daylight to do it, it would mean descending in the dark. We had enough oxygen to reach the top, but there would be none left for the return. Descending requires much less energy than going up, but we would be exhausted; fatal accidents are eight times more common on descents than on ascents. Ask Kami.

Thoughts were swirling through our heads. At a saner altitude, the decision would have been easy, but without enough oxygen, higher cortical centres weaken their control and obvious conclusions can be hard to reach. Mount Everest is a powerful lure. So close to the top, we could feel its upward pull. Two hundred and seventy-four metres. Three football fields. The distance around a New York City block. The chance of a lifetime. Two hundred and seventy-four metres.

Finally: clarity. Our oxygen supply was too depleted to reach the top safely, but not so depleted that we weren't able to realize that. We turned around.

I knew Everest would be there next year. I didn't know whether I would. As it turned out, *National Geographic* decided to continue the project, so I did get the chance to come back the following year. Again, however, I was unable to reach the summit. This time it would be because my expedition turned into a rescue mission when a brutal storm inflicted the worst disaster in Everest's history and I was the only doctor on the mountain.

We should have paid more attention to the comet Hyakutake, when it lit up the night sky over base camp that year. Ancient healers warned that visitations from comets brought death and destruction to those below. They were doubtless thinking about plague and pestilence more than hypothermia and frostbite. Nevertheless, by the time our expedition was over, I wished we had heeded their warnings.

Several major expeditions were on the mountain that year. Besides our *National Geographic* team, there was another American team, a team from New Zealand and a film crew shooting an IMAX movie. After taking a month and a half to acclimatize and to build and supply our own camps up the mountain, we each started off on our summit attempts. The other American team and the New Zealand team were one day ahead of us, the IMAX team one day behind. Four days after setting out, we reached Camp III and were preparing to move up to Camp IV at the South Col the following morning. The American and New Zealand teams had arrived at the Col the previous day and had set out for the summit. Our only link to them was through a radio relay from base camp. We anxiously awaited our scheduled call time to see how they were progressing. When it came, we learned that Rob Hall, the leader of the New Zealand expedition, and five others from his team had summited. We could hear the jubilation at base camp far below. But there were some ominous signs. Five others in the group had turned back early because of high winds, and the summiters hadn't reached the top until two-thirty in the afternoon – a very late time to be starting down. And so far, there was no word at all from the American team. We left the radio on standby, hoping for word that everyone had made it back safely to Camp IV. Our uneasiness grew as the wind blew fitfully, the temperature suddenly dropped and storm clouds rolled over us. Slowly it got dark. The radio was silent too long.

The next report chilled our already-too-cold tent. Caught in the sudden storm, Rob was still up near the summit with Doug Hansen, who was out of oxygen and too weak to descend the Hillary step, a technically difficult rock cliff just below the summit. The rest of the New Zealand team was also having trouble getting back. The American team had all summited, with the exception of its leader, Scott Fischer, who, mysteriously, had not been seen since early morning. Nor was there any word on Makalu Gau, a Taiwanese climber who had last been seen on the summit. Climbers were strung out all over the mountain, some apparently still trying to get down off the southeast ridge. I could visualize every step they were taking. The images of my own summit attempt last year were still vivid in my mind.

That night I lay huddled in my sleeping bag at Camp III,

sipping oxygen, unable to stay warm, or to sleep. As the storm battered my tent, climbers 610 metres above me were exhausted, out of oxygen and exposed to savage wind and cold. They were at the limit of survival. Some, perhaps, already beyond it.

The next morning the reports were confusing and conflicting. Rob was still stuck above the Hillary step. It was unclear what had happened to Doug. The climbers who had managed to make it back to Camp IV before the storm hit reported that it was bitter cold and that the high winds were blowing snow in all directions, creating a zero-visibility whiteout. Climbers had no idea who were in the other tents. Anyone outside the camp wouldn't be able to find it. Eighteen people were still missing, including the entire American team. We heard that two members of the New Zealand team, Yasuko Namba and Beck Weathers, had already been found lying dead in the snow on the South Col.

The wind was blowing snow right through the fabric of our tents, which were flapping so violently that it was impossible to communicate between them, even by shouting. Our two strongest climbers, Todd Burleson, in my tent, and Pete Athans, in the other, had to talk by radio to formulate a rescue plan. They never discussed whether they should go. Their conversation began with how quickly they could get ready. Though I was the only doctor on the mountain, there was never any thought of my accompanying them. Not only would I not be able to keep up, I'd more than likely end up as one of the casualties. Moreover, I had worked with them both for years; I was confident they could take good care of anyone they found alive. But I reminded them to be extremely cautious before deciding that someone wasn't. My presence on the scene would probably not make any difference; treatment options would be limited to the absolute basics. Before bringing any medical supply to this extreme altitude, I had ruthlessly considered its weight and bulk, its usability once it froze, the chance of its being needed and the likelihood that anyone, including me, could function well enough to use it. I gave Todd nearly all the medicines we had – one handful.

In swirling wind and bitter cold, Todd and Pete started up the slope that led to Camp IV. I turned my attention to the medical situation we were in. We still had no clear idea of the magnitude of the disaster, but I knew how thin the margin of survival is near the summit. When I had returned from my own summit attempt the year before, cold and exhausted, I reached inside my down jacket for a drink from my water bottle, only to find that the water, which I had been carrying against my chest to keep warm, had frozen solid. And my summit conditions had been comparatively mild. I was about to learn what happens to human bodies when confronted by the full power of this alien environment.

Remaining at Camp III made sense only if everyone got back safely or no one did. I didn't have enough supplies, or enough room, to take care of anybody there. At 7,315 metres, Camp III was merely a notch cut into the middle of a relentlessly smooth 1,524-metre, 45° ice slope. The platform of ice was no bigger than the tents that rested on it. There was no place to stand outside. Even if I had the medications, anyone who needed treatment would have to be tied in at a 45° angle. Any seriously injured survivors would have to be brought down to Camp II, a much flatter, more open area located at 6,401 metres. There I could set up a makeshift hospital. I listed the medical supplies for each condition I might soon be treating, then contacted base camp, read off my list and asked that the supplies be brought up to Camp II. I didn't yet leave Camp III. I knew that once I did, I would be out of radio contact with the South Col. So I waited.

Todd and Pete made it to the Col. The survivors of the New Zealand team were huddled in torn tents, many too weak or dazed to boil water or put on oxygen masks. The American team had all made it back to Camp IV with the exception of Scott, who was still missing, and his chief guide, Anatoli Boukreev, who was still looking for him. They all had frostbite but were in better shape than the New Zealanders and, since the wind had by this point somewhat abated, were starting down under their own power. A group of Sherpas were starting up from Camp IV to look for survivors and to try to rescue Rob and Doug, still perched above the Hillary step.

Based on Todd's and Pete's assessments, we concluded that none of the survivors were hypothermic or in immediate danger of pulmonary or cerebral oedema. But they were exhausted. The danger of having them try to descend a 1,524-metre ice slope in that condition had to be weighed against the risk of further deterioration overnight at 7,925 metres. On balance, it seemed better to keep them on the Col until morning. We hoped that with rest, food and fluids, they would regain enough coordination and concentration for a safe descent once the weather cleared. The route down the ice slope can be fatal even for a strong climber in good weather, as we had seen the previous year.

While Todd and Pete were recharging the survivors at Camp IV, those of us at Camp III turned our attention to the American team climbers, who would soon be passing us on their way down to Camp II, where they could rest a lot easier. Members of the IMAX film team had come up from Camp II to help us out. They brought supplies and set up a way station in one of their tents to dispense water, food and oxygen, as well as to provide a place to warm up and rest. There was no room to examine anybody in a tent, so I sat by the open door of mine and watched while, one by one, each climber came down the rope and into the IMAX tent. All were exhausted, most had frostbite, but none showed the disorientation or incoordination of cerebral oedema, or the breathing distress of pulmonary oedema. I had a few dexamethasone pills left, which I handed out to those I thought needed them most. Inside the tent, climbers were given a cup of tea, a bowl of soup, some puffs of oxygen and a lot of encouragement. Then they were promptly sent on their way. We were afraid that if they rested too long, they would be unable to get moving again.

The drop in the wind that had allowed the American team to descend proved to be a temporary lull. The storm caught its breath and then roared back more fiercely than before. I began to second-guess our decision to leave the New Zealand team at Camp IV for another night, knowing that conditions on the Col now had to be even worse.

They were. The Sherpas who had gone out looking for survivors returned to Camp IV, forced into retreat by the freezing winds. They had been unable to reach Rob and Doug, trapped high up on the southeast ridge. They had, however, found Scott and Makalu on the South Col, lying in the snow not far from each other. Scott was dead, but Makalu was still breathing, so they dragged him back to camp.

The storm continued to build. The climbers remained hunkered down in their tents all afternoon. Todd happened to look out his tent window into the swirling snow and saw an apparition. Someone was out there. Thinking it was a climber trying to go to the bathroom, he went out to help. The climber staggered toward him, jacket open, a gloveless right hand dangling from his arm. His face was bloated and blackened with frostbite, but it was still recognizable. It was Beck Weathers – risen from the dead.

Todd brought Beck into a tent and laid him down inside two sleeping bags. But no number of sleeping bags would have been enough to rewarm him. Sleeping bags are like thermoses; they retain heat, they don't produce it. Beck had stopped shivering. He was too cold to generate any heat on his own and would have to be warmed from the outside in. Todd and Pete filled bottles with hot water as fast as they could melt snow, and placed them where they knew they would do the most good: in the armpits and groin and around the neck. The arms, legs and head are the parts of the body that 'stick out' farthest from the central core and hence are the most vulnerable to cold. They have large pipes that channel blood into them to keep them warm – a brachial artery for each arm, a femoral artery for each leg and two carotid arteries for the head. The armpits, groin and neck are the entrance points for these vessels. There the skin is thin and there is a large blood flow just below the surface, making them the most efficient starting points to transfer heat into the body. Another way to add heat, though far less effective, is to place it inside the body. Todd and Pete warmed Beck's stomach with tea and warmed his lungs with oxygen. The oxygen would also serve to stoke the embers of his dim metabolic fire.

They were doing all the right things by the time they called me. The only thing I considered suggesting was that one of them take his own clothes off and get in the bag with Beck. A warm body would provide a steady source of additional heat radiating over a large surface. How much heat actually gets transferred through this method is questionable, but I was anxious to rewarm Beck as evenly as possible. I didn't want him to survive hypothermia only to die from being defrosted. His hands and arms and face were severely frostbitten, and possibly his feet as well, which meant that the blood vessels supplying these areas were constricted; any blood trapped inside would be as cold as the outside temperature. Now that we were warming his limbs, that blood would be flushed out into his general circulation. If his body temperature remained low, nerve conduction and muscle contraction would be slow. His heart muscles would be stiff and clumsy. Being hit with a sudden blast of cold blood might be more than enough to turn a coordinated, synchronous heartbeat into a disjointed collection of muscles firing independently.

I was about to suggest that one of them get in the bag with Beck when Todd told me that Beck was coming around. It was too late to carry him down to Camp II that day. He would have to spend the night on the Col, as would Makalu, who was also still alive but, like Beck, in precarious condition. It didn't appear as if anyone else would be coming out of the snow alive. However, Anatoli Boukreev, Scott's chief guide, who had already rescued three climbers, refused to leave the Col. He had been told Scott was dead, but now that Beck had been resurrected, he thought there was a chance for Scott too. He went out into the fierce storm a third time to look for him.

Anatoli's ability to combat the storm was a demonstration of the human body's ultimate performance, of what it is capable of achieving at extreme altitude when finely tuned, fully adapted and highly motivated. Anatoli had been climbing without oxygen. As a guide responsible for the safety of others, he was criticized for this because it made him less strong and less able to tolerate the cold. But as a climber, it meant that he was fully adapted to the effects of low oxygen. A superb athlete, he was even able to outperform many other climbers who had the advantage of supplemental oxygen. He was in the best position to confront the natural environment when disaster struck and the artificial supports had been blown away.

Anatoli reached Scott. All he was able to do, though, was confirm that his friend had lost his battle with the environment. Higher up on the mountain, Rob Hall was losing the battle as well. The Sherpas trying to rescue him had been beaten back by the wind. His companion, Doug, had already succumbed to the cold, and now Rob was alone and out of reach. He could have saved himself earlier had he left Doug, but he refused to abandon the dying climber and now was too weak himself to descend any farther. Risking one's safety for the sake of another, as Anatoli and Rob had done, would seem to be a counterproductive trait in the game of survival, yet it persists in higher animals and humans. A successful species is one that preserves its genes in many individuals, to ensure that some will be passed to future generations. Any one copy is expendable, especially if it serves to preserve the species' genes in other carriers. Altruism may be destructive to the individual but it benefits the species, which is why we value it so highly.

When Rob was told by radio that the Sherpas couldn't reach him, he asked to be patched through to his pregnant wife who was home in New Zealand. He said goodbye and then turned off his radio. He did what he could to protect himself. He crawled into a depression in the ice to block the wind, partly covering himself with snow as insulation against the cold. He took his metal crampons off to slow the conduction of heat away from his body.

That night came the most violent winds I have ever experienced. Our tents at Camp III were anchored to the side of the slope by ice screws. We were sure the screws would pull out at any second or else the tent would shred, sending us tumbling 914 metres down the face. We spent the night fully dressed with our boots on, splayed out over the tent floor, trying to hold the tent down to lessen the pull on the ice screws.

In the morning came word that Beck and Makalu had survived the night and would be brought down to me at Camp II. The New Zealand team, what was left of them, would be able to come down under their own power. We were all grimly aware, though, that it was impossible for Rob to have survived a second night so high and so exposed. The storm had passed and taken Rob's life with it, though the real cause of death was altruism.

Todd and Pete and the group of Sherpas at Camp IV were also responding to the instinct to save their species as they prepared to lead Beck and Makalu down the mountain. I asked Pete to give Beck some dexamethasone before they started. He would first have to melt and warm the frozen ampoules before he could inject one, and then place the other inside his jacket for later use on the road. Beck would need all the help he could get. He couldn't hold the rope in his frozen hands and could hardly even see it, due to what I presumed was snow blindness. They climbed down bunched together, with Pete in front acting as Beck's seeing-eye dog and Todd behind holding Beck's harness to prevent a fall. The Sherpas had left earlier with Makalu. Both convoys were making good progress. It was time for me to start down. As I left the camp I passed Dave Breashears, the IMAX director, and Ed Viesturs, his super-climber film star, on their way up to take the relay from Todd and Pete. I climbed down the 914 metres to Camp II and prepared to receive my patients, who really belonged in an ICU. I would be treating them at 6,401 metres, trying to work out complex medical problems at an altitude where tying your shoes can be confusing.

I commandeered the largest tent in the camp, the New

Zealand mess tent, and took an inventory of available supplies. Everything I had requested the previous day had been brought up that morning, including a heavy propane heater. I also had plenty of help. Henrik Hansen, a Danish climber who was also a doctor, had come up from base camp. Sherpas were eager to do what they could, as were those climbers who still had some energy left. We cleared out the folding tables and chairs, laid foam mats and sleeping bags on the floor, hooked carabiners into the tent frame to hang IV bags and laid out oxygen cylinders and regulators, presetting them for maximum flow. Climbers gathered together two complete sets of dry clothes, and Sherpas went to the kitchen to heat as much water as they could. I systematically arranged my medications and bandages and verified that my medical instruments hadn't been damaged or frozen during the trip up from base camp. Only the IV bags were frozen. The Sherpas took them to the kitchen, thawed them out in the boiling water and returned them to me with a cup of tea drawn from the same water. I hung two IV bags, connected the tubing and left the ends dangling above the sleeping bags. I sipped my tea and waited.

Makalu arrived first, bursting through the door in a bulky hooded down suit, an oxygen mask over his face. He was escorted by a group of Sherpas who immediately laid him down on one of the sleeping bags and began taking off layer after layer of wet clothes. He was wet down to his underwear. His feet had swollen tightly inside his boots and swelled even more once we worked them off. We would have to find a larger pair of boots to put back on him later. He was coherent and only mildly hypothermic, but his appearance was terrifying. He was missing his nose. In its place was a brittle black crust that spread on to his cheeks up to his eyes, which were swollen shut. All his fingers were dark grey plump sausages. His toes and heels were also frozen and grey. There were weak pulses in his wrists and ankles. I marked them with a pen to keep track of his circulation, and prepared to treat the worst case of frostbite I had ever seen.

Most of the damage was irreversible. Makalu had been exposed to high winds and subzero temperatures for over twenty-four hours. With wet clothes, no shelter and no energy left, his body was desperate to retain what little heat it still had. Severe constriction of blood vessels in his arms, legs and face – the worst heat-leaking body parts – had reduced blood flow to nearly zero. With no incoming heat to counteract the environment, the sacrificed parts rapidly cooled to the ambient temperature – far below freezing. The water inside the tissues formed ice crystals, which grew by extracting water from the individual cells. In other words, Makalu's hands, feet and nose were freeze-dried. At this point, frostbite is still reversible. Adding heat melts the water and, like freeze-dried food, the tissues can be reconstituted. With prolonged exposure, however, cells begin to break down. Ice forms within the shrunken individual cells, and as the ice expands, the cells rupture. Endothelial cells, which make up the smooth lining of blood vessels, are especially susceptible to injury. When their surfaces break, the result resembles what occurs when the vessel is cut from the outside. The damage is misread by circulating blood proteins called fibrin, which activate as if responding to a bleeding vessel. The fibrin forms into clots to plug holes that aren't actually there. As the endothelial damage spreads, clots are deposited everywhere, obstructing flow within all the vessels. Even if the body part is warmed, the clots are permanent and flow remains blocked.

It was too late for Makalu's nose, but parts of his hands and feet might still be saved if they were re-warmed rapidly. The decision to thaw him out, however, was not as obvious as it might at first appear. Once thawed, the hands and feet, as well as the rest of Makalu, would have to be kept continuously warm. Blood would be flowing again, through tissues that had been damaged but hadn't fully clotted off. If those parts froze again, they would quickly clot off completely, adding a second layer of dead tissue to the first and leaving Makalu worse off than if his hands and feet had never defrosted. It might be a lot safer, and it would certainly be easier, to preserve them frozen, since the temperature inside the tent was well below zero, and we didn't yet know how we were going to get him down. Thawed-out feet are fragile and will crumble if they have to bear weight. If Makalu had to walk, frozen feet would support him a whole lot better. In treating frostbite at 6,401 metres, the problems are all in the logistics, but I had a lot of help, the right medical supplies, enough fuel and more than enough snow to melt. We could manage the logistics. I would thaw him out.

Into the IV that Henrik had already started, I injected nifedipine, the drug that dilates blood vessels in the extremities. This would bring more blood to the hands and feet but take it from the centre of the body. Given that Makalu was already severely dehydrated, I was afraid that it might cause a precipitous drop in blood pressure, so, having no pressure cuff to monitor him, I periodically placed my fingers over the carotid artery in his neck to make sure the pulsations remained strong. Any weakening and I would have opened wide the IV line and pumped his pressure back up with more fluid. Defrosting Makalu required three tubs of warm water – one for each hand and one for the feet. The Sherpas filled the tubs with water heated in the kitchen tent, and I measured the temperature by dropping a little plastic bubble thermometer card into each one. The water was initially far too hot, but the temperature in the freezing tent quickly dropped it to the ideal temperature of 40°C. Any cooler and it wouldn't be effective. Any hotter and it would burn the skin. Frostbitten hands and feet are incapable of feeling pain because nerves shut down quickly when exposed to cold. The ensuing numbness explains why applying ice is a good pain reliever, but it makes rewarming dangerous. I've seen cases of second-degree burns on top of frostbitten skin because the heating was done too energetically and detected only when the victim smelled his own skin burning. Maintaining the ideal water temperature takes work when the air surrounding it is below zero. As the water in the tubs cooled, it was canted off into empty bottles to make room for the hot water, which had to be added almost constantly. The Sherpas were eager assistants, and though most of them had never used a thermometer before, they quickly got the hang of it, reading the temperature in each tub and deciding on their own when to add water.

Just as we were getting Makalu under control, Beck arrived, led in by Dave and Ed, who had taken the relay from Todd and Pete. I had expected a disoriented, uncoordinated, half-blind, frozen shell of a human, but as he was being eased to the floor, he looked at me and said casually, 'Hi, Ken. Where should I sit?'

Makalu had had the worst frostbite I had ever seen – but that was before I saw Beck. His entire right hand and a third of his forearm, as well as his left hand, were deep purple and frozen solid. They radiated a cold that I could feel against my face even in the subzero tent. They had no pulses, felt no sensation or pain. They didn't even have any blisters, an automatic body response that requires a minimum of coordinated activity between nerves and blood vessels. They were the hands of a dead man, yet bizarrely, Beck could still move them. Hands are designed for both delicate and strong motion and thus are powered by a large array of both fine and bulky muscles. They are such compact machines, though, that most of the muscle mass must be placed outside, in the forearm, connected to the bones in the hand by ropelike tendons. Beck's forearm muscles were still alive, so the tendons were able to pull on his dead bones and produce motion, in the same way strings move a marionette.

Beck had lost his vision on summit day, and I had expected to treat him for snow blindness, which is a common condition in climbers who either lose their goggles or remove them when the vapour from breathing warm oxygen fogs them over. High on a mountain, ultraviolet rays are undiminished by the thin air. When they reflect off ice and snow into the eyes of an unprotected climber, the cornea – the transparent skin window over the lens – gets an intense sunburn and becomes opaque.

Light is blocked from entering the eye, and total, though temporary, blindness results. Like any other sunburn, it is very painful, but it will heal. That healing, however, takes a few days. I was puzzled that Beck was experiencing no eye pain and was able to see me clearly as soon as he entered the tent. Examining his corneas with my ophthalmoscope, I found them surprisingly clear. Each cornea had four scars radiating out from the centre, like the spokes of a wheel. I didn't understand then that they were the reason – not snow blindness – for Beck's temporary loss of vision. He had had a radial keratotomy, an operation in which relaxing incisions are made to modify the shape of the lens so that glasses are no longer necessary. The result of the operation had never been tested at extreme altitude. Corneas get their oxygen supply directly from the air rather than from blood vessels because they have to be transparent. Beck had been breathing supplemental oxygen on summit day. However, since that oxygen reaches tissues only through blood vessels, it didn't help his corneas. Unable to absorb enough oxygen from the air, the corneas swelled, but the scars unevenly restricted their expansion, creating an irregular surface that distorted the entering light. With increasing altitude, Beck's vision became blurrier and blurrier until he could barely see. Now, with the reduction in altitude, his eyes were curing themselves.

We followed the same routine for Beck as we had for Makalu except that Beck's feet weren't frostbitten so a third tub wasn't required. By now the Sherpas had become masters of rewarming, but even with only two tubs to maintain, they were scarcely able to keep up. Beck's hands were blocks of ice and they cooled the water with incredible speed.

Beck was coordinated and fully oriented. As we worked, he talked easily about what had happened to him. He said his vision had progressively deteriorated on summit day, and by the time he reached the southeast ridge he knew he had to turn back. He was overtaken by the storm and, in the whiteout, couldn't find his way back to camp. Realizing that his hands were numb, he tried to warm them by putting them inside his jacket. He removed his right glove to unzip the jacket, but the glove blew away. Though he got his jacket open, he never managed to get his hand inside. Exhausted, he collapsed in the snow.

He said he entered a timeless, dreamlike state, aware of his surroundings but unable to move. A voice intruded into his consciousness when someone leaned over him and said, 'He's dead.' Images and sounds of his home and family floated through his mind, growing more and more vivid and real until they became powerful enough to stir him into action. He got up out of the snow. Reasoning that he would have to face into the wind to get back, since it had been behind him when he set out, he staggered ahead through the whiteout until he became the apparition Todd had seen outside his tent.

Though he recounted all this in a quiet, straightforward way, the story left me awestruck. The windchill factor above Camp IV could only be found on a chart for Mars. Already oxygen-deprived, dehydrated and exhausted, a human body cannot withstand an onslaught of this magnitude for a day, a night and a second day. It was impossible for Beck to have survived. As his temperature dropped, slow nerve impulses and stiff muscles would have left him unable to shiver, dropping his temperature even more. He would have lost the ability to coordinate the large muscles necessary to keep moving, and with it, the heat those muscles would have generated. Too weak to remain upright, he collapsed in the snow, helpless to counteract heat loss while his body temperature began its downward slide toward matching the outside temperature. He would burn the last of his fuel to try to produce heat, but the cold-stiffened muscles of his heart and lungs would be increasingly hard to propel. As his metabolic fire dimmed, his body would begin to shut down. Heart contraction and lung expansion would occur probably only once or twice a minute, their motion so reduced that movement of the chest wall would be almost impercep-tible. So little oxygen would circulate that even his brain, the priority user, would barely have enough to maintain its most elemental electrical circuits. When those few remaining sparks went out, vital functions would stop. That was the sequence that should have begun as soon as Beck collapsed in the snow. Looking at Beck now – alive and talking to me – I felt I was witness to a supernatural event. The contrast between the powerful story and the casual way in which he told it made his testimony all the more stunning.

My human-scale ministrations paled into insignificance. Nonetheless, I had two critically ill patients to keep stabilized through a night in a freezing tent. Our propane heater was kept constantly aimed at Beck and Makalu, and being highly directional, it did nothing to warm either the rest of us or the IV bags hanging from the ceiling. If the bags were allowed to cool, the fluid would drip our patients back into hypothermia. We alternated the exposed bags with others that we kept soaking in tubs of hot water. To prolong hang time, we wrapped chemical hand-warmers around them while they were in the air. I passed the night in the bitter cold of the tent watching the dripping of the IV bags, the flowing of the oxygen regulators and the breathing of my patients. My feet were freezing, though I had put mittens over my socks and then wrapped them in a down jacket. To stay warm I tried to work as much as possible without getting out of my sleeping bag.

An hour before dawn we got a radio message that the helicopter we had hoped for couldn't come. It was still too windy. With no idea when it was going to let up, I had a hard choice to make: keep Beck and Makalu there for another day or try to get them down now. Carrying them down would put them at risk of further injury. Even more, it would mean risking the lives of the rescuers, who I knew were ready to carry out my decision without question. We could try to wait out the wind but at this altitude even simple cuts don't heal; Beck and Makalu, with the injuries they had, would deteriorate rapidly. The need to get them out was compelling. I opted to bring them down the hard way: a combination of walking, sledding and carrying them over the crevasses and down through the icefall to base camp. Rescue teams were organized. We started off, Beck walking with assistance and Makalu being dragged and carried, since he couldn't put any weight on his frostbitten feet.

So intensely focused were my thoughts on how we would get through the icefall, and then how I would manage my patients at base camp, that I didn't notice the wind had died down. I was startled by the noise, and then the appearance, of a helicopter overhead. The pilot was trying to take advantage of the lull in the wind to scoop up two passengers. He touched down lightly on the ice, keeping the rotors going at full speed. The helicopter, and the pilot himself, were both well above their altitude limit. The air was too thin to provide much lift for the rotors or much oxygen for the pilot. If the helicopter couldn't lift off, it could be as fatal for the pilot as a crash. Totally unacclimatized, he would be unlikely to survive more than a few hours.

At this altitude only one passenger could be lifted out at a time. Since he couldn't walk, Makalu was loaded first. The pilot ferried him to base camp, then came back and picked up Beck. He reloaded Makalu in the relatively thicker air at base camp and flew them both off the mountain, pulling off the highest helicopter rescue in history. Beck and Makalu were in a clinic in Katmandu before the rest of us made it back to base camp.

Though left with devastating injuries, Beck and Makalu both survived. Beck lost his entire right hand and most of his left, while Makalu lost large portions of both hands and parts of his feet. Both had to have their noses reconstructed. Their recoveries are stunning examples of the body's incredible ability to endure in one of the harshest environments on earth. Beck's survival, however, transcends the laws of medicine. Beck descended into profound hypothermia – and then, somehow, climbed back out.

OUTER SPACE

GOING OVER THE EDGE

IMAGINE LEAVING HOME AND FAMILY to get inside a camper with five other people. You have worked closely with them for years; some you genuinely like, others you merely manage to get along with. Eating, sleeping, washing, showering and using the bathroom – everything will be done inside this vehicle. Although you have a privacy curtain, you're never more than a few metres from everyone else's sight, sound and smell, or they from yours. You can't open the windows. You can look outside, but all you will find is black sky, unchanging except for the stars. There is no fresh air, no outside sounds, no natural light, no daily rhythm of light and darkness, no weather and no change of seasons. You can't go outside except for an occasional short walk around the vehicle. There are no visitors, no chance for conversations except with the other five inside – and very few of those will be private. There isn't much to do, and yet if you allow your mind to idle, your thoughts will most likely focus on fears of death. Nonetheless, you hope your self-imposed state of isolation will last the full three years that have been planned; anything less would mean sudden death. You're on a voyage to Mars – with no guarantee that it will be a round-trip.

Mounted on a rocket, your camper is a spacecraft. Though roughly the size of a jet airliner, most of the space is filled with machinery, fuel and supplies. When Mars and Earth are at their closest proximity, the outbound journey takes nine months, the return trip nine more. Optimal orbital alignment of the planets will be lost two weeks after arrival, not to return for another year and a half. Since a two-week visit would hardly be worth the trip, you have to be prepared for the longer stay. At times, you will be 402 million km from home.

Astronauts don't have to leave the planet to experience at least some of the physical and mental stresses of living with a small group in a claustrophobic capsule from which there is no easy escape. A training facility that in many ways mimics life on a trip to Mars exists on a coral reef off the coast of Florida. Just 16 km from the animated and popular boating resort of Key Largo, the undersea habitat *Aquarius* sits silently in 19 metres of water. Placed on the seafloor in 1996 and run by the University of North Carolina, it doubles as a research facility for the National Oceanographic and Atmospheric Administration (NOAA) and as a space analogue for NASA. Astronauts are submerged there for weeks at a time to live in group isolation and carry out experiments and training exercises in a programme called NEEMO – NASA Extreme Environment Mission Operations.

The astronauts scuba dive down to the facility and, once inside the dry chamber, breathe pressurized air for their entire stay. This means that their bodies become fully loaded with nitrogen, as any other saturation diver's would be, and they are unable to return safely to the surface without first decompressing for over sixteen hours. Their saturated state does, however, allow them to spend as many as nine hours per day underwater outside their habitat. Since scuba diving makes them essentially weightless, the astronauts have the opportunity to master the skills and perfect the techniques they will need on space walks. Isolation underwater also trains them to focus on complex tasks while maintaining a cautious awareness of the hostile environment surrounding them.

As astronauts venture out along the seafloor, they reel out safety lines from their habitat, which they can follow back if they get disoriented, or lost at night. Small plastic triangles are affixed to the line at regular intervals so that an astronaut can feel the triangle's point and know which way is home. Much like cave divers, they can't surface if they run low on air. Over their heads is an invisible barrier that they can't penetrate without getting the bends because of their saturated state. The safety lines are then cross-connected to form a spiderweb lying over the reef – a grid that can be used to measure rock formations and to determine the precise locations of plant and animal specimens. The oceanographic work is valuable in its own right since it adds enormously to our understanding of coral reefs, but it is doubly beneficial to astronauts, who will use the techniques they learn here to map the surface of other planets.

Along with their surveying work around *Aquarius*, astronauts build elaborate lattice frames under the sea, using the weightless environment to further develop the procedures currently being used to construct the International Space Station (ISS). Sections of the station, sent into orbit in compact form, are opened and assembled like an Erector set. During a mock procedure on NEEMO 5 one day in the summer of 2003, astronaut Clay Anderson was trying to enlarge a hole in a plastic pipe so that he could pass a bolt through it. His knife slipped, cutting into the side of his thumb. Realizing, as he said later, that 'it was probably not a good thing to be bleeding in shark-infested waters,' he and his 'space walk' partner, scientist Emma Hwang, headed quickly back to the habitat. To complicate matters further, the team was scheduled to hold a live videocast that same afternoon with schoolchildren across the United States. The two other astronauts had to handle the videocast, since bleeding in front of a TV camera might have been a bit distracting for the young students. And as if that weren't enough, in the middle of the videocast *Aquarius* suffered a power failure. Though the reserve battery kicked in immediately, the station was now functioning on reduced power. Anderson and Hwang called topside to the Navy doctor on standby, Jay Sourbeer. Dr Sourbeer happened to be on a support boat almost directly above the habitat, having just completed two dives with the NEEMO project director, Bill Todd, the science manager, Otto Rutten and a visiting diver who had been invited by NASA to observe the NEEMO project because of his interest in extreme medicine. The visiting diver was me. Jay and I conferred by radio with Clay and Emma. From their description, we decided the wound would need stitching. NEEMO protocol called for maintaining as much isolation as possible within the habitat; there aren't going to be any guests on a space station. But this was an emergency, and I was a hand surgeon only about 18 metres away. Jay asked me if I wanted to make a house call.

We had already done two dives each, during which I had swum around the habitat – a steel cylinder propped above the seafloor on four stubby legs. The cylinder is about the size and colour of a large school bus, with a box-shaped entry port at one end. Several large portholes allowed me to peer inside as the astronauts ate lunch and did the dishes. I had also swum around the reef, observing the grids and markers, the lattice-work construction and the 'gazebos' – air chambers strategically placed around the site. The divers can enter them to refill their tanks or use them as a safe haven if they have to abandon *Aquarius* in an emergency. Because of our previous dives, Jay and I were close to the maximum amount of nitrogen we could tolerate without having to decompress. We hadn't planned for a third dive, so while we waited for a shore boat to arrive with some additional medical supplies, Otto calculated how much time we could safely spend below. We would have fifty-three minutes.

Jay, Otto and I dove back down toward *Aquarius*, descending past the vibrant reef on which it stands, surrounded by plants and fish of every colour and stripe. I was swimming in an aquarium. Otto directed me to swim up under the entry-port box at the end of the habitat. Suddenly I was out of the aquarium. My head had popped up above a pool of water inside a metal-encased air bubble. A smiling man in a dry T-shirt and shorts was leaning over a railing. He reached out to take my dive gear from me and hung it on the wall – as if we were in a locker room. I stepped up on a metal gate, then on to the deck. 'Welcome to *Aquarius*,' he said. I felt as if I had stepped through a space warp.

We were in the 'wet-porch' – the transition zone from sea to inner space. The air within *Aquarius* exerts enough pressure to prevent seawater from rising up from below, in the same way that air pressure prevents water from spilling out of an inverted glass covered with a piece of cardboard. The result is the 'moon pool', its surface always at deck level, providing open access to the sea. *Aquarius* needs a floor not to keep the water out but to support the legs of air-breathing mammals. There were four of us on the wet-porch now. James Talacek, my official greeter, maintained the generators, compressors and heaters that kept *Aquarius* humming. He instructed me to use the hot shower in the corner to rinse off the salt water. Then he threw me a big fluffy towel and told me to dry myself thoroughly so that I would carry as little moisture as possible into the already too humid air within the habitat. Jay did the same. We wrapped ourselves in dry towels. Otto stayed on the porch to wait for us and count the minutes.

James pulled a hydraulic lever in the ceiling, opening the hermetically sealed hatchway in front of us, and we stepped into a small entry chamber. There were a sink and counter on one side; on the other were gauges, valves, computers and tangles of wire. I was thrust, nearly naked, into the twenty-first century. Standing before me was Clay, our wounded astronaut. He was resting his arm on the counter, his thumb over the sink, where Emma was conscientiously irrigating it with fresh water. Emma stepped aside and I took a look. The lighting was a little dim due to the power failure. James found a flashlight and held it overhead.

For the second time in my life, I was operating by flashlight. I thought of Hermanigildo, the little boy who had sliced open his hand trying to clear weeds from under his father's dugout canoe in the Amazon. Now I was with an astronaut constructing a prototype space station underwater. The two exist at the very opposite extremes of life on our planet, yet their injuries were poignant reminders of how alike and vulnerable we all are.

Clay's laceration, only about 25 mm long, ran obliquely across the last joint of the thumb on the outside. There was no nerve, tendon or major blood vessel damage. Although it would require stitches, the repair would be easy. I cleaned the wound and placed a sterile drape around the hand, then injected lidocaine at the base of the thumb to block the nerves. I inspected the wound from the inside, and once I was sure there was no other damage, I sewed it closed with three nylon sutures and covered it with a compressive dressing. It was a routine repair, but the first operation I had ever done wearing only a towel.

Jay and I escorted Clay back to his bunk, passing through another portal into the main chamber that contained the living quarters. There I met the other two astronauts, Garrett Reisman and Peggy Whitson, the mission leader, who had already logged over six months on the ISS. They had just finished their videocast. While Clay rested, we relaxed for a few minutes, sitting at the dining table, conversing over cups of hot cocoa and looking at the ever-changing parade of sea creatures that passed by the large porthole. Less than an hour before, I had been one of those creatures looking in.

There was a knock on the entry port – Otto was giving us our ten-minute warning. My patient was doing well enough now to escort me to the wet-porch and tell me to 'Stop by any time you're in the neighbourhood.' I submerged through the moon pool, leaving behind a tiny island in a vast ocean, but taking with me the haunting sense of what life is like for astronauts inside a cramped metal canister surrounded by an immense and hostile wilderness. I had been inside for fifty-three minutes. You, on the other hand, on your journey to Mars, will be inside for three years.

Can a group of six humans survive so far away from Earth for so long? In space our enemies would be some of the same environmental hazards we encounter on Earth – radiation, cold, low air pressure. However, there they exist at such extremes that our body's slowly evolved adaptations and carefully learned defences are instantly and piteously rendered irrelevant. Skin pigmentation, sunscreen, tightly woven fabrics and even umbrellas may protect us from solar radiation at the beach or in the desert, but we have never been exposed to the most powerful radioactive particles: the cosmic and solar rays that are blocked by Earth's atmosphere and magnetic field. Once we travel outside those shields, we become fair game for a whole army of high-energy atomic particles that travel at nearly the speed of light and create tiny nuclear reactions each time they pass through our bodies.

We may be able to maintain our internal temperature by shivering and adding layers of clothes in Arctic conditions that would otherwise be cold enough to halt the chemical reactions that keep us alive. But what chance do we have when exposed to temperatures approaching absolute zero, –273.16°C, the temperature at which all atomic motion stops?

The lower the air pressure, the lower the temperature at which a liquid boils. People who live in the mountainous regions of Colorado have to use pressure cookers to maintain enough air pressure so their soup won't boil before it's cooked. At the highest camp on Mount Everest, 8 km up, water boils at such a low temperature that it can be drunk right away. In space, 209 km above the Earth, the boiling point drops below 37°C, meaning that our blood would boil away just from body heat alone.

There is one more environmental enemy in space, and it's not like anything we might experience naturally. Though less deadly than the others it remains all the more diabolical because it subverts an immutable fact of life on Earth: constant gravity. Because gravity doesn't change, no living organism has adapted with any special designs to deal with variations in its intensity. Once we leave the Earth, gravity problems begin immediately – first there's too much gravity and then too little.

Knowing all this, you have nonetheless volunteered for this pioneering expedition. You've gone through all the years of training, and your rocket to Mars has finally lifted off. Accelerating toward a speed of 16 km per second, your body weight is effectively increasing 45 kg per second. Your stomach has gotten so heavy that it's falling through your abdomen, though it's still tethered to your throat by a hyperextended oesophagus. The delicate air sacs at the bottom of your lungs are breaking from the pressure and air is leaking out. Your blood now weighs nine times more than it did several heartbeats ago and your heart is no longer strong enough to pump any of it up to your brain. The launch may be proceeding smoothly, but your body is frantically trying to handle the chaos created by the rapidly increasing gravitational force.

Gravity pulls everything to Earth. When a rocket pulls you the other way, you're swimming upstream. The more it accelerates, the greater the drag on your body and the heavier your body becomes. An acceleration of 3Gs (three times the force of gravity) is necessary to pull away from Earth. Fighter pilots routinely make turns in the 7G to 9G range. Doubling the G force has doubled your weight. At 2Gs your face droops and your internal organs begin to shift. As the third G is added, the liquids in your body (which is 60 per cent water) become hard to lift. They pool in your legs, and your heart has a tough time pumping them up to your head. One more G and your vision dims – you're experiencing a 'greyout' because the blood flow to your eyes has diminished to a trickle. You're at your limit. If no countermeasures are taken and the Gs continue to mount, flow to your eyes will stop completely and your vision will black out, though you will still be able to hear and think. Adding another few Gs will stop blood flow to your brain, rendering you unconscious.

Just as it does in high altitudes, the lack of oxygen affects your whole brain, not just your higher senses of consciousness. Your lower 'body maintenance' areas begin to misfire, sending uncoordinated signals. This results in seizures when your large muscle groups are affected, or convulsions when your internal organs are involved. Mistimed signals to your heart will cause inefficient pumping or even stop it entirely. What blood your heart does manage to distribute around your body will not contain its normal amount of oxygen, for your lungs are also suffering from the increased G force. Your trachea, or windpipe, has been flattened. The alveoli in the lower part of your lungs have broken under the pressure, allowing air to escape before it can be transferred to your blood. Though your upper alveoli are still intact, the increased weight of your blood prevents it from reaching them.

Your body is breaking down everywhere. Your kidneys can't filter enough blood to make urine. The compressive pressure on your spine is so severe that the fluid-filled discs separating and cushioning the vertebrae are in danger of exploding.

Your response to this overwhelming onslaught will be valiant but feeble. Pressure receptors in your aorta and carotid arteries, sensing a decrease in blood flow (just as they would at high altitude), send signals that increase heart rate while at the same time constricting blood vessels. This forces more blood through a narrower space, increasing the pressure head in an attempt to force blood into the tissues. Evolved over thousands of years, the system makes the internal pressure adjustments necessary on Earth, but it was never designed to fend off the six-second, ninefold increase in gravity that fighter pilots experience, or even the threefold increase you are experiencing now.

To survive liftoff, you will need help – anything that works. No treatment is either too complex or too simple. Lying on your back will greatly increase your tolerance of G force. With Gs pushing on you from chest to back rather than from head to foot, the direction of acceleration becomes 'eyeballs in' rather than 'eyeballs down'. Body weight is spread over a larger surface, organs are compressed front to back (the shortest internal distance) and to distribute blood throughout the body the heart only has to pump it sideways. Combating gravity in the horizontal position is familiar to all of us; we apply it regularly when we go to sleep.

Lying down on the job still won't be enough. You should be grunting. Pilots are trained to perform the 'hook manoeuvre': bending forward, taking a deep breath and then grunting the word 'hook'. This closes the throat and increases the blood pressure in a way very similar to bearing down to move your bowels.

You're under attack by increased gravity, a powerful and strange enemy, and your defensive response is to lie down and grunt? That's just not going to be enough. Venturing into space requires a completely different form of treatment – applied technology.

As G forces mount, your blood and other body fluids become increasingly reluctant to leave your legs and lower body. Encouragement will be provided by your anti-G suit – a series of bladders inside your space suit that are wrapped around your legs and abdomen. The bladders inflate instantly in response to the increased G force, much as a car air bag inflates in response to contact. The resulting squeeze injects fluid back into your upper body, and periodic pressure pulses continue to milk blood from your legs. Your blood pressure rises to its normal level, and your beleaguered heart can maintain blood flow. I have applied the same suit many times in hospital trauma units, where it is called 'antishock trousers'. It saves lives by preventing shock from blood loss.

Your respiration is also under attack. G stress will flatten your windpipe and break your air sacs, leaving your lungs incapable of providing adequate oxygen to your blood. The treatment is a positive-pressure breathing device, an air pump that forces high concentrations of oxygen deep into your lung tissues. That reduces the amount of work you need to do to breathe, and each breath brings you more oxygen. Similar units, commonly known as respirators, are used on patients in ICUs to ease the burden on their failing lungs.

The ultimate treatment for too much gravity is not a medical treatment at all but a device called an auto-recovery system. Once you lose consciousness, recovery is not instantaneous even when the G forces are off-loaded. It will take twelve to fifteen seconds for consciousness to return, and then another fifteen or so seconds until you become reoriented and coordinated enough to resume control of a rocket or an aircraft. Rockets can be guided from the ground, but with the high speed and ground-hugging capabilities of high-performance jet aircraft, thirty seconds is more than enough time to crash. Body sensors on an unconscious pilot will activate an auto-recovery system that decelerates the plane and flies it until the pilot recovers. Theoretically, the system can prevent crashes. It has been resisted by airplane pilots, however, who are uneasy about letting machines do their flying.

That uneasiness has been justified on at least two occasions, when Soviet crews headed for the *Salyut 4* space station involuntarily tested how much sustained G force a human body can withstand. Less than a minute before takeoff, cosmonauts Vladimir Titov and Gennadi Strekalov, aboard the *Soyuz 10a*, spotted smoke and flames after spilled fuel ignited around the base of their booster rocket. The automatic launch abort system failed to activate; its control wires had already burned. Fire quickly engulfed the rocket and Titov felt the booster below him sway. The escape system was not crew controlled. The only way to activate it was for two ground controllers, sitting in separate buildings, each to turn a key on their control panel simultaneously. Twelve long seconds later the cosmonauts felt two explosive bolts firing – the capsule had disengaged from the booster. The emergency engine was ignited by radio signal, and Titov and Strekalov were instantaneously subjected to a violent acceleration as their capsule pulled away from the burning rocket. Six seconds later the booster exploded. Their flight lasted less than six minutes, landing them within sight of the burning launch pad. They had experienced 17Gs, but only for five seconds. Neither sustained any serious injuries. They had passed their medical examinations and were drinking vodka long before the fire at the launch pad was out.

The record for enduring the longest sustained high G force is one that cosmonauts Vasily Lazarev and Oleg Makarov would rather not have been a part of. The empty first-stage booster of their *Soyuz 18* spacecraft was supposed to jettison at an altitude of 116 km, with propulsion shifting to the second stage. The parts failed to separate, but the second stage ignited anyway. Powered from the middle, the rocket began twirling end over end like a baton. Ground telemetry was unable to detect any abnormality in the flight path, so it took some time, and some increasingly loud shouting, for the cosmonauts to convince mission controllers to fire the escape rocket. The capsule fell from a height of 193 km after a wild twenty-one-minute suborbital ride, and landed in the Altai Mountains, 1,609 km from the launch site. It tumbled down a slope and was about to go over a cliff when its parachute lines got tangled in some trees. Miraculously, Lazarev and Makarov survived with relatively minor injuries, though the acceleration gauge in the capsule had broken after it reached 20.6Gs.

That's not something you want to imagine as you watch the acceleration gauge in your capsule holding reassuringly steady at 3Gs. The combination of recumbent position, anti-G suit, positive pressure breathing and occasional grunting is enough to counteract the gravity overdose you experience as your rocket heads for temporary Earth orbit. Technology has cured your first space disease, proving that humans can be kept alive under high G loads. With the engine roaring behind you, you're pressed back in your seat but still feeling reasonably comfortable as the spacecraft reaches its orbital speed of 28,967 km per hour. The main engine cuts off. Suddenly you feel yourself being flipped upside down. But there's no problem with the spacecraft – and you aren't really being flipped at all. The problem is in your ears. Having just dealt with too much gravity, you now have to deal with too little. As you enter orbit, welcome to your next space disease, weightlessness.

Gravity has been a part of life ever since there has been life – a downward pull that has been incorporated into our bodies throughout their development, most especially in the design of our body stabilizers, the vestibular organs of the inner ear. There are two kinds: the semicircular canals and the otolith. To do their jobs right, they both need gravity to be unchanging and reliable, but now suddenly inside your ears the gravity is gone. The semicircular canals consist of three tubes perpendicular to each other that detect turning motion. The tubes are filled with fluid, which flows when the head rotates. Hairs connected to nerves stick up into the tube and sway with the current. When the head turns right, the fluid sloshes left; when the head turns up, the fluid sloshes down. The hairs stimulate the nerves, and the signals tell the brain which way you are headed. The otolith is a platform suspended in a gel just above another set of hairs projecting up from a nerve. It floats backward when you move ahead and pushes down when you rise, so it detects forward-backward and up-down motion. You can see for yourself how fine-tuned the vestibular system is. If you pass your finger back and forth in front of your eyes, the image will very quickly become blurred, as your eyes can't keep up. Then hold your finger steady and move your head back and forth at the same speed. The image stays in sharp focus. The vestibular reflexes can react much faster than the muscles that control eye motion to stabilize an image.

Under so-called normal conditions – that is, conditions that have existed for the past 2 million years – the eyes and vestibular system work extremely well together. The eyes are quick enough to keep an animal in focus as it passes across your field of view, while chasing it down means moving rapidly over uneven terrain as your eyes keep the animal in perfect focus so you can throw your spear accurately. Nevertheless, the system does not have the design specifications to deal with G forces, acceleration and zero gravity. The limits of the vestibular system are soon exceeded in an unearthly environment. Though capable of detecting the up-and-down motion of an elevator, it cannot make sense out of space travel. The brain cannot comprehend that gravity has changed that much; it was never programmed to process such unlikely data. Rather, it constructs other 'more likely' scenarios to explain the confusing sensory input.

When the launch began, you felt sudden acceleration because your otoliths were pulled backward to the extreme. When you reached orbit, however, the main engine was cut since the pull of gravity was neutralized. Your otoliths, which had been pinned back by the constant acceleration, were suddenly released and thrust forward. To a brain that has grown up in a 1G world, this can mean only one thing: your head has suddenly gone from tilting way back to tilting way forward. No longer held down by gravity, the otolith has lost contact with its underlying hair cells. Your brain 'knows' this can only happen when you stand on your head, so even as your spacecraft is slipping smoothly into orbit, the brain concludes that you have just tumbled forward and are now upside down.

For a pilot taking off from an aircraft carrier, the design of the otolith and the semicircular canals can prove fatal. As they are catapulted off the deck, pilots receive a nearly instantaneous impulse of 5Gs. The otolith shoots backward and their brains tell them they're tilted up at least 45°. If no outside visibility counters the misinterpretation, a pilot may believe he is climbing too steeply and will 'correct' his flight path by flying into the sea. Furthermore, the semicircular canals were built to let you know you have turned your head; they were not made to orient pilots making prolonged banked turns. At the start of the turn, the tube fluid is set in motion and the pilot knows he is turning. As the turn becomes steady, the fluid 'catches up' and stops flowing. When the pilot finishes the turn, momentum drives the fluid the other way, and he may assume that he has overcorrected. Should he continue to correct this perceived overcorrection, he will progressively lose altitude and put the plane into a spiral dive. The pilot can make an appropriate adjustment when there is a visible horizon, but if there is nothing to be seen out the window, the only recourse is instrument flying. A scenario such as this may well have caused John F. Kennedy Jr's tragic crash into the sea on a foggy night off Martha's Vineyard.

A pilot can disregard what his inner ear tells him when and if he can see that it is not so. The brain will believe its eyes over its ears every time. Consider what happens when you watch a wide-screen movie of a roller-coaster ride. Though you know your chair is not moving, you become nauseated and may even throw up. When there is discordance between visual and vestibular input, the brain puts its trust in what it sees, setting in motion a well-coordinated bodily response to a problem that does not exist.

You cannot change the design of your otolith and semicircular canals, but you can learn to overcome the illusions they create through training sessions in a centrifuge. As part of your training you were strapped into a barbershop chair mounted on a turntable. When the chair was spun and tilted, you experienced all the effects of increased gravity, learning what signals to expect from the balance centre in your cerebellum and how to disregard them by imposing logical thought from the frontal lobes. In other words, you learned to use your brain to override your brain. The treatment is effective but can be frightful. Cosmonaut Viktor Savinykh told me that one of his sessions was abruptly terminated – not by him, but by the scientist monitoring him. The scientist had said, 'I can't take it any more.'

A centrifuge creates an outward pull (centrifugal force) by spinning around a fixed point to which it is tethered. You experience centrifugal force when you're pushed against the door of a car making a high-speed turn. Contrary to what people assume, the weightlessness of a body in orbit around Earth does not stem from the absence of gravity; the distance isn't even halfway far enough away to escape Earth's pull. Orbital altitude means the distance required for gravity to be weakened enough so that it can be exactly offset by the centrifugal force created by orbital speed. The rocket and its contents are still tethered by gravity but precisely balanced by the outward pull of the centrifugal force created as it rotates around Earth. Were gravity to let go completely, the rocket would fly off into space, the same way you would if a car door suddenly swung open.

As you spin around Earth, your body, which only recently weighed 272 kg, now discovers it weighs nothing at all. You release your harness and float upward, experiencing the delightful sensation of being as light as air. But inside your body, things are floating too. Your internal organs have become buoyant. You subconsciously keep tightening your abdomen to push them back down. Having just been jerked back and forth by excess Gs, the poor vestibular system in your ears now receives no input at all. The pressure receptors in your feet aren't firing either. As a result of all this you feel as if you are upside down. Orienting yourself depends on visual cues, but they are often absent or misleading. Nothing in the capsule helps you distinguish top from bottom – up is wherever your head is. Turning a dial provides a sensation that the dial remains stationary and the control panel and entire spacecraft are rotating around it in the opposite direction. Nausea wells up inside you.

Space sickness affects about two-thirds of the astronauts and seems to have no correlation with a propensity for seasickness or even for airsickness. Lots of treatments have been tried, but none have worked too well. Despite not having gravity to work with, however, the vestibular system somehow adapts, and the condition usually clears up within a few days. Until it does, however, a space capsule cannot be a drug-free zone. Temporary relief is provided by medications such as Phenergan or scopolamine, mixed with a little Dexedrine to prevent drowsiness. Sick astronauts and cosmonauts take downers and uppers to keep flying.

You're still in low Earth orbit so that you can temporarily dock at the ISS. Extra fuel tanks have been prepositioned there by earlier missions; these tanks will be hooked up to your rocket for the long voyage ahead. Out the window the space station comes into view. It is either above or below you – the terms are irrelevant – but you have to align and slow the spacecraft to achieve a gentle docking of two vehicles, both travelling at 28,964 km per hour. The approach has to be head-on and the closing speed reduced to less than 3 km per hour. Anything else would cause a collision – and provide a quick reminder of how thin are the walls separating you from the vacuum of space.

Cosmonauts Vasily Tsibliyev and Aleksandr Lazutkin and Astronaut Michael Foale got just such a reminder when their *Mir* space station was rammed by *Progress M34*, an unmanned supply ship. Tsibliyev was handling the docking of the ship by remote control – using two joysticks to align the ship and a lever to brake it while watching a blip on a radar screen. The last part of the docking had to be done visually, but when the ship approached, Tsibliyev, Lazutkin and Foale could not see it out the window. Finally the ship came into view when it passed in front of an extended solar array that had been blocking their line of sight. It was both bigger and nearer than they had expected, and closing on them much too fast. Tsibliyev fired the ship's retrorockets too late to reverse its momentum. They watched helplessly as the *Progress M34*, travelling at 1 metre per second, passed over the docking port and crashed into the belly of the space station.

There was a violent shaking. Foale felt his ears pop, as might occur from the decrease in air pressure when a plane takes off. That he was alive at all told him the impact was not-yet-catastrophic, but the pop meant the cabin was decompressing. *Mir* had sprung a leak. With atmospheric pressure rapidly decreasing, the spacemen's bodies were undergoing the same kind of stress they would have endured in a balloon floating upward into ever thinner air. Diminishing outside counter-pressure allows air within closed compartments to expand. The air-filled cavities within human bodies distend. If the pressure drops too rapidly or too far, they will burst like overinflated balloons.

While the body has outlet valves to equalize pressure in most of its air spaces, their functioning appears quaint compared to the exigencies for survival during rapid depressurization. A valve in the middle ear connects to the throat and can be opened to relieve pressure by swallowing hard – something passengers do instinctively when their flight takes off. Expanding gas in a stomach can be relieved by belching, and pressure in the intestines can be relieved by passing gas. Although there is plenty of air in the lungs, much of it in tiny air sacs far from the outlet valves of the mouth and nose, excess air will work its way out through normal respiration. All these natural escape valves work well when internal gas buildup occurs slowly, such as while hiking up a mountain trail or after eating spoiled food. They were never meant to keep up with pressure changes in leaky spacecraft, however, and they quickly fall short.

Moreover, some body spaces that contain gas have no outlet. All our joints are bathed in synovial fluid to reduce contact friction between bones, much as lubricating oil helps gears mesh smoothly. The nitrogen dissolved in this fluid is held in solution by atmospheric pressure. When the pressure is suddenly reduced, the expanding gas has no natural escape route from the joint. It can only be removed slowly, through metabolism by body enzymes. The decompression of the astronauts on *Mir* was exactly analogous to what scuba divers experience. Survival on board *Mir* depended on slowing the leak so that it would be gradual enough for humans' cumbersome adjustment mechanisms to work and buy enough reaction time for the brain to devise, and the body to carry out, a plan to prevent the pressure from dropping so low that no further adjustment would be possible.

The space station was a construct of interconnected modules. If the ruptured module were sealed off, the pressure in the rest of the station could be maintained. A series of cables and a ventilation tube passed through the module, all of which had to be cut to allow the connecting hatch to shut. Working furiously, Lazutkin and Foale managed to cut the cables and tube, but even with the hatchway cleared, they still couldn't close the hatch. It was hinged on the inside, and they were unable to pull it shut against the stream of air being sucked out the hole. Tsibliyev opened an extra tank of air to slow the steady drop in pressure and prolonged their 'time of useful consciousness' – the time in which they could act purposefully to correct the situation. Lazutkin got the idea of sealing the passage with a hatch cover originally used to cover the opening until the module had been fully assembled. Once the lid had been placed in front of the hatchway, the vacuum inside the module sucked it against the opening, forming an airtight seal. The pressure stabilized, and the men survived.

The pressure drop hadn't been rapid enough for anyone to explode, but it could have been enough to cause the bends – yet, strangely, no one got bent. Many pressure drops have occurred in outer space, albeit in less critical situations, but so far no astronaut has ever been afflicted with the disease. Weightlessness may prevent the bends in some way not yet fully understood.

The *Soyuz 11* mission was heralded as the beginning of a new era in space exploration. Its crew was the first to board a space station, the *Salyut 1*, where they studied the effects of weightlessness. After twenty-four days, they reboarded *Soyuz 11* for their return flight. At a height of 161 km, where atmospheric pressure is still close to zero, they fired explosive bolts to separate their re-entry capsule. The explosion knocked loose a pressure-equalization valve, and the capsule began depressurizing. The limit of human tolerance is about one-third atmospheric pressure, a level not reached until descent to an altitude of 8 km above Earth (coincidentally, that is precisely the air pressure and height at the top of Mount Everest, putting the summit at the very borderline for human survival). The pressure-equalization valve is not supposed to open until 4 km up (the approximate height of the Everest base camp), where the outside pressure of one-half that at sea level is adequate to sustain life. Two of the cosmonauts tried to shut the valve manually but, weakened by prolonged weightlessness, they were unable to turn it fast enough to prevent total depressurization. With no surrounding atmosphere, air is sucked out of the lungs and blood boils at body temperature. After the capsule completed the automatic reentry and landing sequence, the recovery team arrived on the scene. They were surprised to find that the cosmonauts had not yet emerged. Opening the hatch, they found out why. The capsule had survived intact; the three crewmen, still strapped into their seats, had not.

Thoughts of what happened on *Mir* and *Salyut 1* fade away as your craft lines up with the space station and coasts smoothly into the docking port. You and the rest of the crew are focused on the tasks ahead; these involve attaching the auxiliary fuel tanks and verifying vehicle system readiness prior to leaving Earth's orbit. This will require numerous space walks.

Space walks are high-risk exercises because they put astronauts at their lowest level of protection. Once you step outside, your margin of safety is as thin as your space suit – your personal spacecraft. To keep you alive it must surround you with enough counterpressure to hold back the vacuum of space, insulate you from temperatures varying from over –150°C to 150°C above zero, provide you with an oxygen supply and handle carbon dioxide removal, and protect you from any haphazard micro-meteor strikes. Your space suit has to do all that and still be flexible enough to allow you to mount solar panels, install cameras and repair who knows what.

Were your space suit inflated to the same sea-level pressure as your space capsule, you would float out looking something like the Michelin tyre man. Shoulders and hips might be able to move against the high pressure, but the small muscles in the hands would have no chance. Tightening a bolt, much less fixing a telescope or even opening a door, would be hard if you couldn't move your fingers. Cosmonaut Aleksei Leonov made his first space walk outside *Voskhod 2* in a fully pressurized suit. When he tried to get back inside, he found the suit was so cumbersome that he could hardly grab the door handle, and had to perform an emergency depressurization before he could fit through the opening. So suit pressure needs to be reduced. One-third sea-level pressure is a good compromise: low enough to provide some flexibility yet still high enough to hold back the space vacuum. But then slipping into a space suit from a fully pressurized cabin would be like going from sea level to the summit of Mount Everest with no acclimatization. And we've learned how the body reacts to that.

To acclimatize safely, your decompression – or reduction to space suit pressure – should take at least four hours, and that's far too long to wait around for each space walk. Astronauts cut preparation time to forty minutes by modifying the environment in the cabin and space suit to narrow their differences. Less difference means quicker adaptation.

Cabin pressure was dropped to two-thirds of an atmosphere just after docking, so your body is already starting to adjust while you're performing other tasks. Hours later you get into your suit, at one-third atmospheric pressure, and start breathing pure oxygen. Though there's less air, there is enough oxygen in it to keep your lungs working to capacity. The mini-acclimatization and the supplemental oxygen work in space as well as they do on Everest.

The moment arrives. Standing in the open hatchway, you launch yourself into the void by pushing off on your toes. The feeling is eerie and overwhelming, and you need to adjust to the sensation of being in slow motion. You float over to do some maintenance work. All workstations feature a handhold or foothold for stability. If they didn't, pushing a button would propel you backward off the rocket, and tightening a bolt would spin you counterclockwise into space. Both of these things you manage to do, and many other things as well, but the work is strenuous, and doing them in a space suit is like exercising inside a thermos. The problem of heat buildup is addressed by water-cooled underwear. Interwoven into its fabric are hundreds of yards of tubing through which water is pumped before being passed over an ice pack. Water-cooled long underwear is available for especially hard work.

Modular units that need servicing or that are most likely to need repair work are designed with round corners and easy access. Still, sometimes unanticipated repairs have to be made. Astronaut Jeff Hoffman related to me how fixing the Hubble Space Telescope (whose myopia was the biggest blunder in the history of optics) required his team to work at some very awkward angles and to squeeze into compartment corners that were never meant to be entered by astronauts. Sharp edges and pointy instruments can puncture a space suit more easily than a micro-meteorite. Should something happen, however, sensors will note a pressure drop and automatically inject high-pressure oxygen from a backup canister, which can feed a 6 mm leak for half an hour. That should be enough time to get back inside. If not, the rapid depressurization would theoretically cause the bends. Should it ever happen, the treatment would be to put the astronaut into a functioning space suit, block the relief valve and then pump the suit up to twice normal pressure. The suit would act like a recompression chamber. Better not to get bent, though. Besides being very painful, the treatment ruins the suit.

Attaching fuel tanks, solar panels, cameras, antennas and other assorted equipment and then visually or manually verifying each system takes several sorties of teams from the spacecraft and the space station, and in various combinations. Finally, though, your work at the space station is done. It took either three days or forty-eight days, depending on where you were standing. On Earth, the planet made three spins past the sun. On the station, orbiting at 28,967 km per hour, the sun rose and set and rose again every ninety minutes. This rhythm is hopelessly confusing to your pineal gland. Located at the base of your brain, the gland had been accustomed to gently secreting the sleep-inducing hormone melatonin in response to the gradual onset of darkness. In space your sleep pattern is interrupted – in fact, you're not sure you've slept at all since launch.

Your biggest impediment to sleep, however, isn't the pineal gland's confusion; it is you. Like all astronauts, even veterans, you're so excited about being in space that you don't feel tired and don't want to waste time sleeping. This is your last day (or last sixteen days) in Earth's orbit, and though you're scheduled for a few orbits of sleep, you would rather look out the window. Being exhausted, however, you are unable to resist visual cues. Each time the spacecraft goes into darkness, you fall instantly and soundly asleep. You awaken bright and full of energy when it comes back into sunlight. You go through an entire sleep-wake cycle every ninety minutes. To get some solid sleep, all you have to do is pull down the shade – except that out the window, rolling under you, is a gorgeous aqua-blue planet. Soon it will disappear from view as you head toward Mars. These few remaining orbits may be the last you will ever see of your home.

The main engines power up for the trans-Mars propulsion. Thrusters fire, reorienting the rocket toward outer space. You feel the pull of gravity once again; orbital equilibrium has been broken. This time it's a gentle pull as Earth slowly releases you from its final hug. You turn away from the window, now filled by the blackness of space, and look around the cabin: your home for the next three years.

Through a combination of complex machinery and clever contraptions, life in the spacecraft can feel almost like home. Cabin pressure is the same as at sea level, and the air is cleaner than on Earth. The temperature remains a constant and pleasant 21°C; your clothes are appropriately casual. A refrigerator keeps drinks cool. Food is preserved by dehydration. The menu consists of 120 choices. A day's meal might include scrambled eggs for breakfast, a ham-and-cheese sandwich for lunch, and shrimp cocktail, steak and broccoli au gratin for dinner. There are thirty beverages to choose from. Hot and cold tap water is plentiful because, conveniently enough, the fuel cells that combine hydrogen and oxygen to produce electricity give off water as a by-product. There is no dining table, but with adhesive straps you can attach a serving tray to your legs. Despite zero gravity you can eat your food with ordinary silverware, so long as there are no sudden starts and stops. There's a watertight shower stall for periodic, and mandatory, use, though sponge baths are more practical. Every two days you put on fresh socks and underwear, sealing the dirty laundry in airtight bags until you put it through the washing machine. The toilet is very similar to earthbound ones except that air current rather than gravity draws the waste into sealed containers. To prevent hair pollution, the men have power shavers with vacuum attachments. The women keep their hair in nets so it won't get caught in machinery, though for photos and videos they're allowed to let it flow freely, thereby creating the ultimate 'big hair' picture.

You have all the comforts of home, except that your survival depends entirely on an artificial environment maintained by an array of machinery too complex for any one person to understand or maintain. This home is a tiny oasis encased in a metal shell sailing through the vast and deadly realm of outer space. Lingering as it does in the back of your mind, this thought never lets you fully relax. A sudden noise gets your immediate attention, like the loud bang astronauts Jim Lovell, Jack Sweigert and Fred Haise heard in *Apollo 13* as it was on its way to the moon. They very quickly determined that two of the three fuel tanks had exploded. Lovell looked out the window and saw gas escaping from a hole blown out the side of the command module. Most of the supply of power, oxygen and water was gone. Three hundred and twenty-two thousand kilometres from home and headed in the wrong direction, they would need four days to circle the moon and return to Earth. 'Houston', Sweigert famously radioed to Mission Control, 'we have a problem.'

A body can survive a few days without water but only a few minutes without oxygen. The supplies in the *Apollo 13* command module were lost, but luckily the lunar landing module contained more than enough oxygen and just enough power and water to get them back. Had the fuel tank been the tiniest bit stronger and exploded on the return flight, the lunar lander would have been depleted, leaving the astronauts with no lifeboat.

The most sinister threat to the three astronauts' survival was their own breath. They had enough oxygen, barely, but soon they had too much carbon dioxide. The natural by-product of respiration, carbon dioxide is poisonous in high concentrations. The problem is neatly taken care of on Earth, where plants take it in and produce oxygen as their by-product. But a space capsule is not an ecosystem. Canisters of lithium hydroxide are needed to 'wash' the carbon dioxide out of the air. There were only two in the lunar lander – not nearly enough to absorb the exhalations of three astronauts for four days.

In a sealed room, carbon dioxide will kill you long before you run out of oxygen. Breathing is stimulated not, as might be expected, by too little oxygen in the blood but by too much carbon dioxide. The respiratory centres in the brain that monitor both gases are far more sensitive to carbon dioxide concentration. As its level rises, your breathing rate will increase, to blow off more of the gas. At toxic levels, your breathing becomes very rapid and shallow, too inefficient to get adequate air into the lungs.

To minimize their carbon dioxide output, the *Apollo 13* crew did their best to limit physical exertion. Nevertheless, the lithium hydroxide canisters, functioning on reduced power, were saturated after a day and a half. Warning lights flashed. There were extra canisters in the command module, but their square pegs wouldn't fit into the round holes of the lunar module filtration system. Through trial and error, engineers and astronauts on the ground improvised an adapter, using only items that would be available on the spacecraft. Following step by step the instructions radioed up to them, the *Apollo 13* crew assembled a jury-rigged contraption of plastic bags, cardboard, parts scavenged from a space suit and a lot of duct tape. An hour later they were breathing easy – and so was the ground-support team. Without that low-tech solution to a high-tech problem, the mission would have ended in tragedy. Instead, *Apollo 13* became 'the most successful failure' in NASA's history.

NASA is justifiably proud of its performance, but that doesn't provide much assurance to you and your fellow Mars travellers. Actually, the story is disquieting. A similar crisis on this mission would most likely not have a happy ending. The *Apollo 13* explosion occurred a mere 321,860 kilometres from Earth – a distance that can be travelled in four days by a spacecraft and one allowing almost instant communication with Earth by radio signal. An explosion on the way to or on Mars, depending on interplanetary alignment, could occur anywhere from 161 to 402 million kilometres from Earth, with a return transit time of six to eighteen months. Radio communication might have as much as a forty-minute delay. In an emergency, survival will be up to you.

A sobering thought as you head out into deep space. Right now you've got a cold and a backache and your bladder is full. Your stouthearted shipmates, equally puffy-faced, can't seem to get comfortable and are getting up to pee a lot. Liberated from the bonds of gravity, the water within you no longer feels the need to stay in place. Since your body is more than 60 per cent water, major shifts in fluid distribution are occurring. You've got 'bird legs'. The 2 litres of water normally held down in your legs by gravity has floated upward, causing overload in the head, especially in the nose and sinuses. The congestion causes a loss of taste and smell, and this makes you feel as if you've caught a cold. Free of downward pressure, the stack of vertebral bones that make up your spine has drifted apart, allowing the fluid migrating through your body to enter and swell the cushions, or discs, that separate the bones. The hydraulic pressure cranks your spine out, making you 51 mm taller than you were on Earth. Sounds great, except that the sudden pull gives you a backache. No position feels comfortable. You feel as if you need to stretch your muscles, but generating the necessary counterforce is not so easy; pushing against something in a space capsule just moves you away from it. Inside the major blood vessel of the chest – the aorta – and of the neck – the carotid artery – receptors monitor the amount of blood in the body, relaying the information to the maintenance centre in the hypothalamus. Not understanding the idea of weightlessness, your brain interprets the extra volume flooding the receptors as an overload. It signals for the release of hormones that reduce whole body fluid by filtering out more through your kidneys and into your bladder.

So you're an intrepid explorer with a stuffy nose and sore back who frequently needs to go to the bathroom. You're also still suffering from bouts of space sickness. Yet incredibly, after several days in space, you adapt. Your body begins to function smoothly on reduced fluid volume, draining swollen tissues and relaxing stretched muscles. Sinuses clear, back pain vanishes. Having reached a new fluid equilibrium, you no longer have the urge to excrete excess water. Your nausea and dizziness have subsided as well. The vestibular system has a built-in ability to reset balance – otherwise we would be unable to adapt to our own growth or weight gains. But that kind of adjustment happens at a leisurely pace and is dependent upon feedback from gravity; it won't happen fast enough for a space flight. Therefore the brain reprogrammes itself using visual cues. You have learned to play by a new set of rules.

Emboldened once again, you and your small group of earthlings are ready to encounter your most formidable interplanetary enemy: the nuclear army waiting in ambush as soon as you venture beyond Earth's protective fortress of atmosphere and magnetic field. Deadly radioactive particles and waves come from two sources – cosmic rays and solar radiation.

Cosmic rays originate outside the solar system, generated by the explosions of supernovas and other giant stars, ripping apart atoms of hydrogen, helium and iron – the primary constituents of the stars – and releasing protons and alpha rays, highly energetic subatomic particles that radiate outward at nearly the speed of light. Coming from every direction in the galaxy, they weave a thick fabric that penetrates all of 'empty' space. The other source of deadly radiation is closer to home – the sun. Sunshine is the result of an ongoing thermonuclear reaction, one by-product of which is the release of protons and electrons that stream outward in a steady flow called the solar wind. Once in a while a sun storm erupts, creating a solar flare that can rise 1,609 km above the surface and release protons at 10 billion times the rate of the solar wind. The blast is unidirectional, can last for hours or days and its timing is unpredictable. Travelling close to the speed of light and aimed toward a rocket travelling from Earth to Mars, it would take about two days to arrive.

Cosmic rays and solar flares are two galactic villains to which earthbound creatures have never been exposed. The menagerie of deadly particles and rays that they produce is kept at bay by our planet's environmental shield. Earth is surrounded by a gas-filled atmosphere and ringed by a two-part magnetic field called the inner and outer Van Allen Belts, both of which originate and terminate at the North and South magnetic poles. Many space particles carry electric charges, and as they approach the frontiers of Earth, some are drawn into, and captured by, one of the two magnetic belts, creating two rings of intense radiation. Some of the trapped particles ride the magnetic field to the North Pole, where they descend through the atmosphere, reacting with oxygen to create a fluorescent green glow known as the aurora borealis, or northern lights. The atmosphere promptly absorbs particles that somehow manage to filter through the Van Allen Belts. As these particles collide with atoms of oxygen, nitrogen and, especially, water vapour, they lose energy and eventually fizzle out before they reach the ground. The only waves that make it through this obstacle course are the visible spectrums that light up our planet, and the ultraviolet rays that give us our suntans.

Inside the space capsule, radioactivity can't be seen or felt, but the reading on the radiation gauge indicates a sharp rise. At 805 km above Earth you enter the inner Van Allen Belt; 9,656 km later, the reading drops, only to rise again at 24,140 km and then slowly taper off over the next 16,093 km. You've exited the outer Van Allen Belt. Now you are outside the bars of Earth's protective cage, vulnerable to the relentless bombardment of cosmic rays and the unpredictable bursts of solar flares. Every particle that collides with your body will trigger a nuclear reaction. A high dose will generate enough explosive energy to break down molecular bonds in your cells and kill you outright. Smaller doses will alter strands of DNA, the highly sensitive and complex genetic molecule present in all cells that controls cell duplication. Loss of that control in reproductive cells will lead to genetic malformation. In the rest of the body, it will lead to the chaotic proliferation of abnormal cells known as cancer.

How much radiation damage you will incur on this mission is unknown, but it will be far higher than on any previous flights. Only the moon landings took astronauts outside the Van Allen Belt, and those missions were of relatively short duration. Space shuttle flights and space station operations happen below the Van Allen Belt, so the radiation level remains low. An exception was the mission to repair the Hubble telescope, which had been placed in a high orbit so that it would be free of as much atmospheric interference as possible. The astronauts were exposed to radiation in the Van Allen Belt, roughly the equivalent of two chest X-rays per day, a high dose. Nonetheless, until this mission, flight times for astronauts had been short enough so that total exposure to radiation was less of an occupational hazard than for airline attendants, who work at a lower altitude but spend much more time in the air. As for solar flares, they are easily visible to an astronomer, who can give a two-day warning to deorbit a spacecraft or evacuate a space station. Plenty of time.

But you're on your way to Mars. The cosmic rays will add up, and you won't be able to outrun solar flares. Putting full shielding on the spacecraft would have made it too heavy to lift off. Partial shielding would have been far worse than nothing at all, for when cosmic rays penetrate shields, they interact with the metal, forming highly radioactive neutrons. Because neutrons have no electric charge, their energy is not easily deflected. They have a high affinity for hydrogen, one of the two elements in water, and water is the primary ingredient of human beings. This is the reason why neutron bombs would kill people so effectively yet would not disturb buildings.

The cumulative effect of long-term exposure to cosmic rays may be unknown, but the effect of a solar flare is easy to calculate: one dose is fatal. So intense is the burst of radiation that you would die within a few hours. You need a safe haven when a solar flare is moving toward you, and your spacecraft provides one – a small, well-shielded shelter into which you and the crew can retreat for the hours or days it takes for the intense radiation to pass by. As for the cosmic rays, space contains no water to absorb them like Earth's atmosphere does, but your spacecraft boasts a plentiful supply, stored in hollow tanks surrounding the sleeping quarters and workstations and affording at least some protection from the steady bombardment. How effectively this works you won't know for years. If you do return from space, you will only be able to say, 'I survived the trip to Mars . . . so far.'

Months pass. Radiation may be the biggest environmental threat to your mission, but the greatest danger of all may come from within. Echoing the words of the cartoon character Pogo, you may one day radio back to Earth, 'We have met the enemy, and he is us.'

Powerful human forces are put in play when you seal six people into a container for three years. The enormous blue globe that was turning outside your window a few months ago is now but a speck in a black void. Everyone you've ever known intimately, everyone who truly understands you lives on that speck, moving 40,233 km farther away every hour. Your family wanted you and needed you at home, but they stood behind your decision to depart on mankind's greatest adventure. You left them, some would say forsook them, to adopt a family you need to get along with for the next three years.

That family consists of a highly select group of pilots and scientists whose qualifications for inclusion were not what you had first imagined – otherwise you might not have been selected yourself. Six hard-charging test pilots with alpha male personalities would have been a recipe for disaster. Nor can the group have six leaders. Members were picked to blend commanders with conciliators, introverts with extroverts, talkers with listeners. The group is multinational, co-ed and ethnically mixed, as befits any enterprise of this historic magnitude, not merely for geopolitical expediency but because different perspectives create a stronger team. Besides, six people with the same cultural and social background on a claustrophobic three-year voyage would be incredibly boring – and therefore very dangerous.

Boredom may be the most deadly disease you will face. Its symptoms begin slowly and subtly, the first signs being a loss of energy and motivation. Soon you're fatigued and irritable, and gradually you lose your tolerance for others. You may withdraw and sink into depression or become aggressive and, at the extreme, hostile. Much has been done to prevent, diagnose and treat boredom. Potential crew members were analysed and evaluated through endless interviews and personality tests, then mixed and matched until, like a patchwork quilt, the pieces seemed to fit together to form a harmonious pattern. The interior of the living quarters was humanized. Sharp edges and exposed metal were kept to a minimum. Illumination was subdued, except for reading lamps and the brightly lit multi-coloured instrument panels. Walls were painted in soft hues, and large windows were placed in positions easy to look through, though other than the background of stars, there's nothing out there to look at.

You have far less contact with the outside world than a prisoner. Stimulation has to come from within, but by this point, systems maintenance and experimental protocols no longer take up much time or inspire much interest. You have the very latest in entertainment and educational technology – exciting at first, but it all got stale pretty quickly. You were encouraged to bring along hobbies, but how many times can you look at your, or somebody else's, stamp collection?

The key to treating boredom is to keep stimulation dynamic and personal. With an up to forty-minute delay in communication, you might be able to play a chess game with someone on Earth, but you can't hold a meaningful conversation. Though you can occasionally receive video broadcasts of your family and hear their voices live over the radio, dialogue is mainly limited to e-mails; these take days to complete. You try to take part in family life, even help manage family affairs, but the frustrating delays make it harder and harder to stay in the loop. Events on Earth are passing you by, decisions are being made without you. You feel more and more irrelevant and isolated.

Ground control does what it can to keep you in touch with the planet. They beam news at you, arrange continuing education courses and even encourage you to vote with absentee ballots. But earthbound institutions seem increasingly less important. Your world is here, in this space capsule.

You don't even have night and day. It is always dark outside the window. No diurnal rhythm to regulate your biological clock. This was a problem I experienced, though for the opposite reason, when I spent a month in Antarctica. There during the summer the sun remains constantly overhead, like a naked lightbulb in a windowless room. With no daily light change, melatonin has no idea when to begin its nightly hormonal flow. I slept when I felt physically tired or when I had nothing else to do. I awoke when it was time to resume work or when I felt fully rested. I carefully recorded each day's events in a log, yet when I returned home I found that my log consisted of only twenty-three entries; I had been gone thirty days. Deprived of any outside cues, my body had reverted to an average thirty-six-hour day – awake about twenty-four hours for each sleep of about twelve. It was responding to some inherent need for replenishment rather than to the exigencies of survival for a human, a light-adapted animal that hunts and gathers during the day and seeks safe shelter at night.

The environment inside the spacecraft doesn't promote anything resembling a normal sleep pattern either. You're zipped into a sleeping bag tied into the wall so you don't float away. Without downward pressure, you don't feel the reassuring contact you were accustomed to on Earth. Machinery noise is incessant. Crews work in shifts, so a light is almost always on somewhere, and somebody's bound to be moving around. Dozing off as necessary may work for a month in Antarctica, but it won't fly in space for extended periods of time. Here, sleep has to be carried out on schedule, like any other part of the mission, no matter how badly you may want to stay up past your bedtime. You use sleeping pills to restore order, but they don't induce good-quality sleep, and there are unpleasant side effects. More than once you've awakened confused and disoriented, wondering where you are. Or were.

Your body is deteriorating. You've lost muscle tone, your bones are weak and your blood volume has dropped. Muscles and bones designed to hold the body up against the force of gravity are unemployed. Without steady work, any muscle will atrophy, losing both substance and strength. Bones not subjected to continual stress will become osteoporotic, losing the calcium framework needed to bear weight. The excess calcium leaches into the blood, whose volume is already diminished by all that urinating you did to compensate for the fluid that sloshed up from your legs when you first became weightless. The calcium gets concentrated when it passes through the kidneys, coalescing into hard balls. When a ball gets trapped in a filtering duct, you experience the sudden and excruciating pain of a kidney stone.

Other, less noticeable changes are also under way. Your decreased blood volume has concentrated the oxygen-carrying red cells in the blood that remains. Not programmed for this possibility, the body reads the increased concentration as an excess of cells rather than a lack of fluid. It resets the proper haematocrit by destroying red cells, making you anaemic and tired. Your heart has less blood to move around, what it does pump has no weight and the demand for output is greatly reduced given that you do heavy lifting with one finger and can cross the entire spacecraft by pushing off on a single toe. Your heart has it pretty easy, and like all other muscles, it weakens from lack of use.

Your immune system, your body's protection against disease, also weakens. No one understands why this happens. The phenomenon, which occurs in Antarctic encampments as much as on space stations, seems, strangely, to be caused by any prolonged isolation in close quarters. Outer space, however, also introduces some unique infection risks. Germs from an unconfined sneeze, unimpeded by gravity, will spread out in three dimensions and hang suspended in the air. Transported from Earth in and on our bodies, our germs may no longer be relatively harmless. Like you, they have been subjected to space radiation and weightlessness. They have most likely undergone frequent mutations and evolved through thousands of generations, adapting rapidly in order to ensure their own survival – not yours. The once-benign bacteria and viruses may by now have been transformed into virulent carriers of unrecognizable diseases.

The changes to your heart, skeletal muscles and bones do not represent a chaotic breakdown of your body's systems. Rather, they are the purposeful, coordinated responses of a body trying to maintain equilibrium in a strange new environment. Living in space requires only occasional exertion, so the heart weakens. Were conditions more rigorous than normal, the heart would strengthen. Spacemen barely need to walk. However, they do a great deal with their upper bodies. Hence, while they lose muscle mass and bone stock in their legs, this is not the case with their arms – much like people with paralysed lower limbs. These are appropriate adaptations that the body will fight to the death to maintain.

Nonetheless, a weak heart, flaccid muscles and thin bones, while perfectly compatible with life in a weightless spacecraft, won't be adequate once you reach Mars. Your resilient body will not readapt to a life of gravity and increased physical activity nearly fast enough to prevent you from being incapacitated once you touch down. Any treatment to reverse your adaptation to space will probably be doomed to failure because it would attempt to override the body's most fundamental drive to survive. However, without some countermeasures while you're in space, your body will crash on Mars even if the rocket lands safely.

It's your turn for your daily two-hour workout on the treadmill. The small compartment it's in smells like a locker room. The last person using it didn't towel off very carefully; a large glob of sweat floats above the machine. You know whose sweat it is by the smell. By now you believe you can recognize whose it is simply by looking at it. You're fed up with that astronaut slob.

And with all this exercise. You put on your harness and attach yourself to the treadmill with elastic bands so you don't fly off backward with your first step. Other elastic bands span your knees, hips, elbows and shoulders, increasing resistance across those joints. Lately you haven't bothered to attach them.

They make the workout much harder, and you're already tired. No one will notice. They're probably doing the same thing. Still nearly a month to go before you reach Mars – plenty of time to build your body back up. The exercise is boring; you quit after an hour and fifteen minutes.

Not that you need the extra time. This momentous trip has turned into monotony. At first everything was new and there was lots of work to be done. Daily instructions from Mission Control used to be exciting; now they're irritating. Lots of things are getting irritating, such as the silly puzzles they periodically have you solve to measure mental and physical agility. And the weekly questionnaires. How many times can you check boxes about your mood, about what you think of your fellow astronauts, about what you think they think about you? Doctors at home have no idea what it's like. The co-pilot's laugh is so annoying. The systems engineer keeps asking you for help when you just want to be left alone.

Besides being monitored by humans, machines watch over you. Sticky pads and transducers on your skin signal your pulse, blood pressure, respiration rate and galvanic skin response – the ability of your skin to conduct electricity, which increases when there's some salty perspiration on it, as would occur when you're nervous. A sensor on your tongue tests your saliva for cortisol, a hormone whose level increases during stress. A pill inside you is transmitting your body temperature – you need to swallow a new one each time it passes out of your digestive tract. Some of these systems I field-tested on the NASA-sponsored expeditions to Mount Everest. There, climbers were grateful to have real-time sensors that could detect physical changes in their condition and possibly save their lives in dangerous situations. Here in space, where you feel relatively safe and comfortable, these readings are more likely to reflect emotional changes, hence the monitoring feels like a continuous lie-detector test.

The most intrusive machine of all is the face camera each of you has at your workstation. It records your facial expressions, matching the pictures with preloaded images of your various moods. The camera is programmed to detect any changes in the pattern or frequency of your expressions that might indicate emotional problems. It's supposed to act as an early warning system. As far as you and the others are concerned, however, it's an individual surveillance camera. More than once you have thrown a towel over the lens.

Such an act of rebellion wouldn't sit well with ground controllers, except that they never find out. Questionnaires and performance tests are relayed back to Earth, but patterns detected by the face camera, combined with input from body sensors, are fed through an onboard computer, which analyses the information and responds with appropriate advice. Depending on the magnitude of the problem, anything from a few words to an entire visual programme will appear on the screen to engage you in an interactive dialogue. You know you're talking to a machine, yet the instant treatment can be very effective. The computer isn't designed to work out unresolved childhood conflicts; it's simply trying to get you past a stressful situation. Each time you cover the camera lens with a towel, you get a programme about controlling your temper. This actually makes you feel ashamed, and you take the towel off. One time you even apologize.

Machines are ready to help with physical problems as well. A sensor is on board that can scan your entire body if you are sick or injured, and create a three-dimensional image that will be beamed back to Earth. Doctors there can rotate, enter and magnify the image to identify your problem and prescribe treatment. If surgery is indicated, a surgeon on Earth can do it virtually, directly on the computer image, free to try different approaches and techniques individualized specifically for your condition. Once satisfied that he's performed the best operation possible, he'll programme it into the computer and beam it back so it can be downloaded into a robot. Like a player piano, the robotic surgeon will carry out your operation – efficient and flawless, though perhaps lacking style and compassion.

The system worked well in trials on Earth but, all the same, you'd rather not see how it works in space.

Your spirits lift as you approach Mars. Your tolerance and sense of humour return; you feel yourself regaining mental equilibrium. There's renewed energy throughout the spacecraft. The crew – with one exception – seems to be coming out of hibernation, busily preparing for the Mars orbital capture now only days away. Space passengers turn back into astronauts. The exception is the doctor. He's been spending all his time in his bunk, unwilling to treat anybody for weeks. His computer psychotherapist hasn't helped. Neither have the messages from home, or the advice he's received regularly from Earth doctors after he flunked that questionnaire. He takes some medications on his own plus some others you are told to give him; nothing seems to have much effect. The crew remains uneasy about his condition, but in such a confined space a formal meeting without him is impossible, so you've been reduced to whispered discussions in twos and threes when you think he's asleep or on the treadmill. Everyone hopes he'll be drawn out of his depression by the pull of Mars.

The red planet looms large outside the window, its colour all the more intense after months of space blackness. So different from the blue planet you left long ago, it's awesomely beautiful and at the same time repellent, even terrifying. Though you're sure the computer has detected your fear, you don't share your feeling with the others, to whom it would only seem counterproductive. Instead you smother it with thoughts of the momentousness of where you are. Then you narrow your focus to help with the task at hand – deflecting the rocket into Mars's orbit.

The orbital insertion proves flawless. Dead ahead is the Mars entry transition station, waiting for you for four years. It was launched during the previous period of favourable planetary alignment – along with a ground station containing a nuclear power plant. The plant has been functioning on the surface, manufacturing oxygen and water out of the hydrogen in the

Martian atmosphere. Sensors have already signalled Earth that enough breathable air and fuel have been stored for the mission's return voyage. Were this not so, your manned mission wouldn't have been launched in the first place.

Rendezvous and docking with the transition station proceed smoothly. The station consists of two modules – a rotary engine to be attached in orbit and a descent vehicle to guide the capsule to a Mars landing. The landing won't happen for another three weeks. Until then you will undergo preadaptation so that you can make the transition to the 0.4G of Martian gravity. The treadmill exercises, even when they were carried out strictly (which, at least in your case, they weren't), only slowed the deconditioning that comes with zero gravity; they didn't prevent it. You have been weightless for six months. Your muscles are weak and lazy; your bones are as thin and as prone to fracture as those of a postmenopausal woman suffering from osteoporosis. Reversing those changes will take weeks of intensive exercise and also require the introduction of artificial gravity.

After a series of space walks, the rocket fuel tanks have been separated from the capsule and the rotary engine has been attached. The engine is turned on. The capsule starts to spin, creating centrifugal force. Within the capsule you feel a pull toward the outside walls that simulates gravity. Once again there is up and down. Objects have weight. Your body dutifully adapts one more time. The adjustment proves easier here than it would be on returning home, since Martian gravity is less than half of Earth's. Still, the augmentation has to be done gradually, so the capsule spins slowly at first, then a little faster each day.

There isn't enough time to wait passively for your muscles and bones to strengthen. The treadmill goes full-time now – everyone feels the need to build his or her body back up. But not the doctor. He thinks he's fine the way he is. You try to convince him otherwise in a long discussion, during which you tactfully inform him of the crew's genuine concern over his physical and mental health. The session ends abruptly when he says, 'Tell them I'm okay or I'll kill you.'

The physical effects of prolonged flight seem easier to cure than the mental ones. The rest of the crew feels fit, anxious to get to Mars, though they know full well that should they have a problem, there won't be any rescue team or rehab programme. As a last precaution, everyone drinks chicken soup. The high salt content draws water into the blood, counteracting the loss of blood volume. Besides, chicken soup will cure whatever ails you.

In preparation for Mars entry, the rotator is detached and the descent module connected. The fuel tanks will be recaptured and refilled for the return journey. They'll be left in orbit for the next year and a half. There's no place to leave the doctor.

The descent rockets fire and the capsule drops out of orbit. Mars comes up from below, spreading out under the window in ever finer detail: broad waves of red desert gouged by deep ravines and peppered over with black boulders. The ride down is surprisingly short. Retro-rockets fire, the capsule hovers momentarily, then settles down with a gentle thud. You've landed.

Now the capsule becomes your Mars habitat. The increased gravity is more than you experienced while in orbit so you allow a day to get used to it. The sun sets and rises. Everyone feels impatient to get going. The first priority will be to verify the on-site supply of fuel and oxygen. This will require only a short walk; the capsule landed very close to the nuclear plant. You're scheduled to be the first one out.

You don your space suit, step into the lockout chamber and wait twenty minutes while the air pressure gradually drops. A light goes on to tell you decompression is complete. You release the lock on the outer hatch and pause for a long moment. You're collecting your thoughts. Finally, you push the hatch open and the chamber fills with dull red reflected light. You're now enveloped in Martian air, although you don't feel any difference inside your space suit. The square hatch opening has become a picture window with an eye-level view of a vast, rust-coloured desert. The harshly lit rocky rubble casts sharp black shadows against the hard flat ground: a silent, motionless, lifeless world. You put your head outside. The desert expanse reaches beyond sight in every direction, broken only by the incongruous foreground presence of the prefabricated nuclear plant and its oxygen silo.

Placing your hands outside, you pull forward to bring yourself up to the edge of the hatch frame and find you are taken aback by the resistance of your own weight. With a second effort you overcome your inertia and stand poised on the threshold. Your foot comes down heavily as you place it in the first toehold in the side of the ship. Weakened muscles make your legs hard to control. You take two more steps down to the little platform just above the surface, then turn to face outward, as befits a pioneer about to take the first step on Mars.

You are standing on a platform atop twenty thousand years of technology, the culmination of an evolutionary process that began on Earth 4 billion years ago. The first, simplest life forms, competing for survival against each other and against the environment, relentlessly redesigned and refined themselves into myriad combinations, some complex enough to conceive of and then construct their own creations. Making use of their inventions, humans have expanded into ever more extreme environments. Of the 15 billion humans who have ever lived, you were precisely positioned in the line of evolution to be the first living thing to try to survive beyond the planet.

You step off your Earth-built platform to make contact with Mars. The hard ground seems to come up too quickly against your surprisingly heavy foot. You feel a snap. A searing pain rips through your calf, and you twist and fall to the ground. You know beyond doubt that you have broken your leg.

Lying in the red desert, you're gripped by the cold feeling that you may never see Earth again. Mars may be the altar on which you are to be sacrificed, a victim of high-tech hubris. No matter how many wondrous inventions you've brought along for your well-being and protection, your body has proven too vulnerable. Its exquisite ability to adapt was never designed for export. Even though your man-made defences have not been breached, the mission to Mars has already become overwhelming. Can your broken leg be fixed? Without a doctor? If so, will it heal? And what about that doctor, with whom you'll have to survive for the next two and a half years? You've only just arrived. What next? How adaptable can your made-on-Earth body be when it's placed on another planet? How far and how long can your will to survive carry you? One hundred million miles and three years from home, have you fallen over the edge?

CONCLUSION

THE WILL TO SURVIVE

CAN A HUMAN BEING WITH A BROKEN LEG, lying alone in the Martian desert, ever hope to see Earth again? What are his chances?

First he must drag himself back into the space capsule. Although he's wearing a bulky space suit, he should be able to manage it; gravity on Mars is much less than on Earth. His bones, weightless for months, will be brittle; the break would most likely require the delicate surgery of open reduction and internal fixation to place plates and screws directly on the bone – and run the risk of further splintering it. It would be a difficult case for the doctor, even assuming that, despite his bizarre behaviour, he's still functional. If he's not, his crew-mates could obtain advice from doctors on Earth. The round-trip delay between each question and answer, however, could be as long as forty minutes – and for the astronaut doing the surgery, this would be his first case. Perhaps the robotic surgeon could do the operation, but the technology is finicky and has never been tried in these conditions.

The advice you receive might well be to amputate your leg. As barbaric as it sounds, it's a much simpler operation. Even if the novice surgeon or the robot somehow manages to get the job done, tight confinement increases your risk of infection, and decreased gravity would most likely delay your healing. The broken leg could instigate an overwhelming infection such as gangrene or pneumonia, or induce a blood clot due to your prolonged immobility. You might recover, but the scenario might also end with you waiting to die, draining the resources of an expedition that would be better off if you died quickly.

Pitting the human body against the planet Mars would seem to be a pitiful mismatch, but throughout this book we've seen, and in my personal experience I've witnessed, many seemingly hopeless contests between humans and hostile surroundings in which the human has prevailed. Given roughly the same intensity of environmental attack, why do some people survive where so many others perish? What factors added up to survival for Antonio and Hermanigildo in the Amazon; Steve Callahan, the Robertsons and the Baileys on the high seas; Mauro Prosperi and Pablo Valencia in the desert; Bernie Chowdhury underwater; and Beck Weathers on Mount Everest? What would lead that crippled astronaut to live or die? I have observed four additive forces at work in the struggle for survival, three of them obvious as they are being applied, the fourth and most powerful one evident only by its effect.

The first factor is knowledge. Antonio's understanding of the jungle and Steve Callahan's mastery of sailing provide the ultimate proof that knowledge is power. Both held within them the wisdom of those who went before – whether tribal elders or veteran sailors – to which they added their own lifetimes of experience. The tools of their survival were in their brains.

The second factor is conditioning. Mauro Prosperi, lacking any desert teaching, survived in an extreme environment because his body was prepared. Rigorous training had created a body capable of responding to stress with peak performance. The gradual introduction of that body to the desert gave it the time to reach its full potential. Likewise, Anatoli Boukreev outperformed everyone else during the rescue attempt on Everest. Because he was a superb athlete who had been climbing without supplemental oxygen, his body was maximally adapted to the brutal environment. Contrast the situations of Prosperi and Boukreev with that of the shipwrecked sailors who were suddenly and unexpectedly tossed out of their comfortable boats into the sea.

The third factor is luck. Everyone needed at least some good fortune, but these determined survivors were able to reduce the amount of luck they needed and increase their time to find it. The only one who was especially lucky was Hermanigildo, who managed to find a fully equipped hand surgeon in the middle of the jungle.

All the survivors had some combination of these three ingredients, but none of them would have lived through their ordeals if they didn't also have the fourth: the will to survive. Sometimes, will alone seems enough to get you through – as it was for a seventeen-year-old girl with little jungle experience and no forewarning who was literally dropped into the Amazon in a miniskirt.

Julianne Kufka, the daughter of German zoologists, was flying to her parents' research station to spend Christmas 1971 with them in the Peruvian Amazon when her plane exploded. She awoke in a kapok tree, still belted into her seat. The jungle canopy had cushioned her 3,048 metre fall, but there was no sign of her mother, who had been sitting next to her, or of any other passengers. Search planes buzzed overhead, but Julianne couldn't see them through the dense foliage – and they couldn't see her. She would have to depend on herself to get out alive. 'I tried to block out as many fears as I could,' Julianne later said, 'so I could concentrate on getting through the undergrowth.'

Setting off in her purple knit miniskirt and carrying her macrame purse, the teenager soon stumbled across a stream and followed it downhill, hoping it would lead to a river. Rivers are the Indians' roads through the jungle, and along a road she might find a traveller. She drank clear water from fast-flowing streams and waterfalls and ate only some candy she had in her purse. She slept sitting up, hunched forward over her folded knees, afraid that animals would be attracted to her body heat and the scent of blood from her insect bites.

Mosquitoes attacked her relentlessly. Some carried eggs on

their bodies, laid there by botflies. The fly's eggs penetrate human skin through the mosquito bite and quickly develop into larva worms. The bite serves as an airhole through which the worm pokes out its head to breathe. If the sores are covered with body fluid and the worm's head is submerged, its respiration will produce air bubbles. 'I could actually feel them crawling and feeding on tissue and fat,' Julianne recounted. 'My whole body was alive with parasites.'

Julianne removed a ring from her finger, shaped it into a hook and used it to extract the inch-long worms. She twisted out thirty-three of them. There were more she couldn't reach.

She kept up her strength by thinking of her father, resolving that he would not be left alone. 'He's not going to lose his daughter,' she told herself. 'I will stay alive.'

After nine days, Julianne was found by an Indian hunting along the river she had been following with such determination. Her will – not her skills, her adaptations or her luck – had kept her alive.

A mini-skirted teenager is not generally the image that comes to mind when we think of a hardened survivor. Yet the will to prevail must exist in all of us, for we are all descendants of survivors. In societies where life has gotten comfortable, the will to survive remains latent. It is perfectly possible now to cruise through life without ever taking a survival test. Ensconced in social services, public health and safety laws, and police protection, we may never be seriously threatened by hunger, exposure, disease or attack. Even when somebody does encounter a survival test, the standard for passing has often been lowered. Dealing with extreme cold might mean nothing more than returning to the ski lodge. A desert wanderer might be able to end his ordeal with a call from his cell phone. The idea of a survival test becomes a mockery when failure means nothing more than getting dropped from contention on a TV show.

One of the few settings where the ability to survive is still critically analysed and tested is in the training of military commandos, such as Navy SEALs. Though candidates have already been preselected for strength, stamina and intelligence, the programme completion rate is only about 50 per cent. Instructors say the successful candidates tend to be the quieter ones who possess the inner strength to keep their bodies and minds functioning beyond exhaustion. One Navy SEAL instructor told me that at the end of 'Hell Week,' a gruelling final exam of physical and mental exercises with very little sleep in between, the sailors are allowed to collapse on the beach. He then says to them, 'Okay, everyone up for a 16 km run.' There's no run, but it tests to see who has the spirit to go on. The ones who stagger to their feet are accepted as SEALs.

Of the 6 billion people in the world today, very few are commandos. The population expanded exponentially once humans organized into civilizations. Without that development, our numbers would most likely still be less than one billion. Probably at least five out of six people alive today depend on society's support and protection for their existence. To realize how dependent we are on civilization, consider this: in a wilderness setting, losing your eyeglasses might well be fatal. Any slight disability, any inherited or acquired disease, would quickly eliminate you from the competition. Who among us are the ones who could make it on our own?

With natural selection so disengaged, the predisposition to prevail becomes randomized and obscured in the population. Some people will retain it, others will not. One who did retain it is Aron Ralston, a twenty-seven-year-old mechanical engineer who set out in the spring of 2003 on a day hike through a remote canyon in Utah. Trying to slip through a 1-metre-wide slot between rock walls, he pushed against the side of a boulder – it shifted, pinning his hand underneath it. He was trapped, but he didn't panic. 'It took some big thinking to calm myself down', Ralston recalled. He coolly tried all his options: he waited, but he hadn't told anyone where he was going and he didn't have that most modern tool of survival, a cell phone; he tried to chip away at the rock with his utility knife; he tried to lift the boulder, using his climbing rope to rig a series of pulleys. The 363 kg stone wouldn't budge.

Night-time temperatures dropped below zero, and by the third day he was out of food and water. He said he went through periods of depression, but 'the majority of the time I was focused on pursuing one of the options.' By the fifth day, he realized that his only chance for survival was to exercise his last option – amputate his arm. Though his pocketknife was far too dull to cut through bone, by levering his arm he was able to snap both his forearm bones above the wrist. Then he applied a tourniquet – and cut his arm off. It took about an hour. He said his strength to endure the pain came from a higher power. 'There was a greater presence than just me in that canyon.'

Even after he had amputated his arm, Ralston still had to rappel off an 18-metre cliff and walk 10 km down the canyon before he was spotted by hikers and rescued by a search helicopter. At the point of rescue, he was only 2 km from his pickup truck. He almost could have driven himself to the hospital.

Testing the capacity for survival doesn't necessarily require being placed in harsh surroundings. All of us, whether living at an extreme or protected by an advanced society, are surrounded by emotional and mental, if not physical, obstacles. We spend most of our lives on the near side of those barriers, even as we long to surmount them. We take the easy way out, arguing, often with much validity, that to do so is safer or more practical. If, however, we gather the will to cross over the obstacle, to confront the emotion or solve the problem, we gain strength from it. Telling the truth, making the sacrifice, doing the job though there won't be any recognition for it – these are mental exercises that strengthen will as much as physical exercise strengthens muscle. Such efforts actually form nerve connections in the brain that make it easier to overcome an obstacle the next time. Moreover, if we persist in the face of adversity, we often get a 'second wind,' much as an endurance runner does when he refuses to quit. The unexpected energy can carry us much farther than we ever thought possible and allow us to triumph over seemingly insurmountable odds.

I've witnessed the powerful effect of will, and the lack of it, not just in extreme settings but in hospitals and homes, where life-and-death struggles are just as real: the terminally ill mother who survived months longer than expected because she wanted to see her daughter's wedding; conversely, the elderly but seemingly healthy husband who died shortly after his wife's death because he couldn't live without her. Whether a girl who survives a plane crash can also survive the jungle will be determined by her ability to call forth whatever mental strength she's developed in her civilized life.

Willpower does seem easier to summon in harsh environments, where the terms of the struggle are brutally clear and there's no room for dependence on others. The winning strategy for survivors would seem to be a contradiction: they're able to focus narrowly and sharply on the demands of their environment while at the same time maintaining their focus on a goal that transcends their circumstances and gives them a larger reason to survive. For Marilyn Bailey, it was to build a new life with her husband; for Lynn Robertson, a religious conviction that her mission was to 'get the boys to land.' Bernie Chowdhury and Beck Weathers triumphed by vividly recalling their families; Andrew Hughes was driven to the limit in an attempt to save his. The motivation is not always for a higher purpose – Pablo Valencia derived his strength by imagining the pleasure he would have knifing the partner who betrayed him.

Even with the best survival techniques, and the highest of goals, the formula won't always work. My friend Rob Hall, an expert mountaineer motivated by a desire to save a failing friend and inspired by a phone call to his pregnant wife, was unable to overcome that brutal storm on Everest. Rob was not so much a victim of the environment as of his genes. He could have outrun the storm if his priority had been only to save himself, but he was a human trying to save another of his species. Sacrificing oneself for the good of the species is pervasive in the animal world – we've seen the striking example of poison-dart frogs. But in higher animals the behaviour requires a conscious decision. The cerebral cortex must override the powerful instinct for survival with an even more powerful signal of altruism. That signal is most intense when the genes in peril are the most similar – a parent ready to die for a child. It remains very strong within families and strong within ethnic groups, lessening gradually as genes become more dissimilar. A species-wide nongenetic version also exists – evident in a soldier's loyalty to his country, or in an individual's willingness to die for a principle.

Six hundred and ten metres below Rob Hall, caught in the same storm, was another of my friends, Beck Weathers. Beck was lower down on the mountain than Rob, but nevertheless exposed to wind and cold incompatible with life. Oxygen-deprived, dehydrated and exhausted, he collapsed in the snow. Helpless against the elements, his vital functions would slow, then stop. The sequence should have been inevitable. Beck lay in the snow for a day, a night and a second day. His frozen body did shut down, but his mind did not. His life was reduced to thoughts – of his home and his family. He refused to die. Thoughts contain electricity. Strong thoughts contain more electricity. Beck's will to survive generated enough power to re-energize his body, get him up from the snow and walk him out of the storm. But how and from where did that will arise?

To explore the source of the will to survive, we must take a journey into the human brain – a 1 kg blob that generates a mere 25 watts of power. That's barely enough to turn on a dim lightbulb, yet the brain contains more unexplained forces and uncharted territory than anyplace else on earth.

Humans aren't the fastest or the strongest species; we've prevailed because we are the smartest. We've evolved the most complex brains, building on the same structures that lower animals possess but adding additional layers and innumerable interconnections in response to the demands of our surroundings.

The most primitive part of the brain is the brain stem, located at the bottom. Like any other stem, it's where the roots come together – the roots being the nerves that branch into every part of the body. The nerves send electrical signals to the brain stem that provide constant updates on internal information such as heart rate, respiration and blood pressure. External information, derived from nerves to our senses, enters the brain through a switching station called the thalamus. If the thalamus interprets the input as danger, it sends a warning signal to the amygdala – the emotion and fear centre of the brain stem – which sounds the alarm, activating the body's defensive reactions. Signals are sent back down the brain stem to increase heart rate and breathing and either to stand stock-still, or move muscles to begin an attack or effect an escape. The system is automatic and operates on an unconscious level. It is based on an ancient blueprint, and it's equivalent to the entire brain of a reptile.

While the brains of reptiles have remained more or less the same for millions of years, the human brain has evolved dramatically from its ancient template. The biggest change has been the development of the cerebral cortex – the stuff that looks like a cauliflower and envelops the more primitive parts. It's responsible for the higher signal processing necessary to evaluate sensory input more accurately and generate more precise and appropriate responses. This process takes time, so the reptile's system still has its uses – a fast reaction may save a life. But now the thalamus switching station has another place to send the signal. The cortex can override the initial reaction once it refines the sensory input, searches its memory and determines that the curvy thing you just jumped away from wasn't a snake but only a stick. The cortex can then send a signal to the amygdala to stop sounding the alarm, and your sense of fear will vanish instantly.

Since higher animals possess more cortex, they can process

input more accurately, and consequently exhibit more intricate behaviour. Each portion of the cortex is responsible for monitoring or controlling one function, such as vision, hearing or movement. The sections must be linked, of course, so that physical actions are coordinated; otherwise a bird could look in one direction and fly in another. But in most animals, no matter how complex the linkage between the parts and how intricate the behaviour, the effect of an outside stimulus will be automatic. Bears hibernate when it's cold, and antelope flee when they see a lion. These unconscious responses to threats from the environment, though complex, are survival instincts.

The behaviour will be routinely predictable unless some system overrides the signal. The most highly evolved animals have shown dramatic expansion of the front part of their cortex, the frontal lobes. This is a tertiary processing centre, taking refined information from each section of the cortex and adding input from the centres that control memory and emotion. The frontal lobes evaluate each component and create a balance that is unique to each individual. They are able to alter prewired instinctual behaviour. The evolution of this capability was the dawn of thinking.

Thinking requires a mind, a software programme that fits into the hardwiring of the brain. The mind occupies the same space as the brain and is bound to its topography. Thoughts arise from specific locations and require energy to form. That's why thinking can make you tired, but it's also the key to mapping the brain with sensitive machines such as positron emission tomography (PET) scans and magnetic resonance imaging (MRI) that precisely localize and measure internal energy levels. If you hear a noise, your auditory cortex lights up. If you recognize the noise to be that of a lion, your memory centre lights up too. And if you've learned that lions are dangerous, your fear centre lights up as well. Energy flows between the centres and on up to the frontal lobes, where information is synthesized. Frontal lobes develop options and make relative value judgements, but they don't make decisions or supply motivation. If we relied on our frontal lobes alone to guide us, we'd be like the proverbial donkey who was equidistant between two bales of hay and starved to death because he couldn't decide which one to eat.

Decisions require someone or something to take control and initiate action – a prime mover. The thinking involved leaves trails that can be followed. Energy flow can be traced from the frontal lobes to a small area lying deep within the cortex that lights up when decisions are made – at which point the flow then reverses, and a cascade of electrical impulses is sent back to the cortex, setting in motion all the functions needed for survival. This signal-generating area is called the anterior cingulate gyrus, or just cingulate, and it appears to be the brain's commander-in-chief.

The cingulate is the decision maker and motivator that enables humans to act in a purposeful manner. Without the cingulate, the body languishes in a state of placid indifference. Consider the example of certain stroke victims, in whom the loss of blood supply to the brain has temporarily knocked out the cingulate. They become incapable of physically responding to the external stimuli that they receive. Once they recover the ability to communicate, they report that, although they were alert and aware, they didn't feel any need to react. If they had heard or seen a lion coming, they would have recognized it and been afraid but wouldn't have cared – a state not compatible with prolonged life. The cingulate may be the key to finding the origin of the will to survive.

Though it arises as a barely perceptible glow in the deepest recesses of the brain, the effect of will on survival is pervasive, dramatic and sometimes powerful beyond our comprehension. It's at work all along a continuum of environmental insults, large and small – anything that challenges a body venturing outside its familiar habitat, from horrific snowstorms to distasteful meals.

The bug-laden bowl of spaghetti served to me in the jungle represented in its own little way a challenge to my survival. The image of a mass of long white strands covered with black dots was transmitted from my eyes to my thalamus switching station. Not seeing an urgent threat, the image was directed to the visual cortex for further analysis. It was confirmed to be bug-laden spaghetti, and the information was passed up to the frontal lobes. There, my memory centre was tapped, and it reminded me that I don't eat bugs. My emotion centre added its two pence by generating feelings of nausea. My information package was complete. Were I at home, I'd have thrown the food away. But a second, competing package had formed as well. This one provided input from my brain stem telling me my blood sugar was low, so I was hungry. My senses reminded me I was in the jungle, and my memory added that the bugs weren't poisonous. My cingulate weighed the two options and chose to favour my survival; it directed me to eat. It then signalled the motor centre of the cortex to coordinate the activity. For good measure, the cingulate also sent a counter-signal to my emotion centres to suppress my nausea. The jungle setting had summoned in my brain enough will to change my behaviour.

At Antonio's hut, however, I could not take a single bite out of a boiled rat head with its hair still on. In that instance, powerful input from the emotions took command of the cingulate's efforts to manage the options of hunger or satiety. Had I been starving, I might have been able to summon the additional will required to partake, but survival was not at stake and my will was fully suppressed by my revulsion. I could eat the spaghetti but not the rat.

It's no surprise that emotions can easily override logic. We see examples of it every day in ourselves and others, like my patient who broke his hand when he found a ticket on the windshield of his car and punched the parking meter. Emotions can prevail over reasoning because the pathways from the amygdala emotion centre to the cingulate decision centre are very well developed. They carry more signal traffic, and thus more influence, than the less developed pathways that go in the other direction. Emotional impulses can overwhelm the capacity for reasoning far more easily than logical control can be exerted over primitive responses. Evolution, however, is a work in progress: higher animals are evolving more connections from cingulate to amygdala, and therefore more control over emotions. Nevertheless, even in humans, the amygdala often still rules.

The plane crash high in the Andes that stranded the Uruguayan rugby team for months forced the survivors into a struggle between their frontal lobes and their amygdalas. Facing starvation, they lived by eating their dead teammates – an excellent food source according to frontal lobe analysis, and the obvious choice for any cingulate. Cannibalism, however, generates a high-intensity signal of revulsion from the amygdala. It's an adaptation, like altruism, that favours survival of the species over the individual. It would be difficult to organize a society if its members always looked at each other as potential meals. People who are able to generate a counterimpulse to squelch the amygdala are those whose will to survive is strong enough to override cultural and emotional considerations. The same tormented decisions determined the fate of the shipwrecked crews of the *Essex* and *Mig* and the snow-trapped settlers in the Donner Party.

For each person in each of these cases, the mind made a decision as to whether or not the organism wanted to survive. If the will is there, the brain initiates actions that are appropriate responses to the environmental stress. Sometimes, though, the brain initiates a response that is 'inappropriate' but more favourable for survival. The mind is able to abrogate the laws of physiology and command responses that seem to break the rules – to 'cheat' in order to aid its own survival.

A pain in the neck is the annoyingly predictable result of any cool breeze that reaches me through an open window at home. Yet I spent a month exposed to the frigid winds of Antarctica and my neck didn't even notice. In that extreme environment, I subconsciously rejected any thought that would have compromised my ability to function. How is this possible? What's the trick here? From what we've seen so far, the answer would seem obvious: some signal is sent to suppress the pain centre. Except that there is no pain centre in the brain. Pain signals are received in the same way as touch or temperature, except that the readings fall outside the body's safe range. Pain input is nothing more than a reporting of facts, but the body has to react to the data quickly to avoid damage. The surest way to do this is to create an unpleasant sensation, a distinct second step that requires stimulating both attention and emotion centres. Unless both are activated, the body won't feel any conscious need to respond. Emotion centres have thresholds dependent on the nature of the individual (e.g., tough guys are more stoic, sometimes) and on cultural considerations (e.g., Sherpas never complain that their feet are cold). If the level of pain activates the emotion centre, the signal gets passed on to the frontal lobes where it will compete with innumerable other signals for the brain's attention. The cingulate will then determine the appropriate strategy and evoke the proper protective behaviour. For me in subzero temperatures, the answer was easy. My brain had more than enough input to know it was extremely cold. My attention centre was fully occupied with the problem, making appropriate behavioural and internal body responses to keep me alive. At home, the awareness of a cold neck would motivate me to close the window or get a scarf, but here it would not change my behaviour, so my brain rejected the pain before it reached the conscious level where it would become a distraction. In the hostile climate of Antarctica, I literally didn't have time for the pain.

Pain can and will break through to the conscious level if behaviour needs to be altered to improve the odds of survival. If the incoming data is way outside the body's safe range, the thalamus will automatically split the input, diverting one signal to the motor cortex for an instantaneous protective response while sending a second signal to the frontal lobe to produce a more modulated but slower response. Your first impulse upon handling a hot cup of tea that is burning your fingers might be to drop it. Your second impulse, which factors in environmental conditions and acquired knowledge, might be to hold on to it, since spilling it will ruin the rug your wife just bought, which on balance might be more risky for your survival than a burned fingertip.

Culture, upbringing and environmental conditions will wire the frontal lobe in a unique pattern that determines an individual's response to extreme stress. Nothing will raise your pain threshold and focus your frontal lobes more than a life in the jungle. That's why Hermanigildo, after slicing his wrist with a bad swing of his machete, could remain calm even though his pain receptors were firing wildly. His eyes were receiving and transmitting a vivid attention-getting image made fearful and gruesome in the visual cortex by additions from the amygdala. His thalamus was madly processing signals trying to keep order. If the emotion centres were to overload, fear would take over and generate a short-circuit – the coordinated but counterproductive response of panic. The cingulate, already well aware of the problem, chose to ignore the pain readings and sent interference signals to the amygdala to squelch the fear alarm.

Impulses generated in the cingulate have such mastery over the brain that they can easily transform perception into reality. In stressful situations, the cingulate regains control of functions otherwise relegated to the body's 'autopilot' – the autonomic nervous system. Incoming signals from the autonomic nervous system monitor functions such as body temperature and hydration and reach no higher than the brain stem, where they are processed automatically. New signals are then routed back to the effector organs, which make corrections by opening or closing heat-dissipating blood vessels close to the skin, provoking sweating or creating a sensation of thirst. Connections to the conscious mind exist, but they are normally activated only by emotion. Being tempted with a cool drink will increase your thirst; becoming embarrassed – a situation perceived as a threat – will increase blood flow (the basis for blushing); sudden fear will increase perspiration (precooling your body in preparation for action); and being attracted to someone of the opposite sex will make your heart pound. Energizing the cingulate with input from the amygdala is as easy as being controlled by your emotions.

Controlling your body responses with thought is a lot more difficult, but it can be learned. Both the ancient art of yoga and the modern technique of biofeedback teach the conscious brain to regain control of the primitive functions it long since relegated to automatic systems. A brain possessed of such control is able to will the body to respond in a pattern not predicted by the 'rules'. Yoga and biofeedback require a great deal of training and practise, but life-and-death exposure to extreme environments seems to facilitate the same sort of communication along those lost channels. Conscious thought can transmit a response that will favour survival, in effect overriding the dutifully relayed signals from the receptor organs. The mind will thereby create a 'reality' that exists only in its imagination but has enough power to command the body to respond in real terms, like soldiers following the orders of a deluded general.

The delusion, as well as the effect, can be subtle, as it was for me in my oppressively hot jungle tent when I misinterpreted a zipper noise as the opening of my tent flap. Soon afterward I felt a cool breeze that didn't exist but nonetheless lowered my skin temperature and stopped my perspiration. Such subconscious control can cool skin in the jungle or heat it in the mountains, as with the Sherpa who came to my tent after falling through the ice on a subzero day and had warm feet. His cortex had overridden the automatic response to shut down blood flow. It had chosen to sacrifice some body heat in order to preserve the limb.

In higher doses, delusions can have more dramatic effects. As the mind separates further from its true environment, it will override clear sensory warnings and respond dangerously to a reality that doesn't exist. Victims of severe hypothermia have reported experiencing an overwhelming sensation of warmth. Mixed with confusion and disorientation, it can lead to the phenomenon of 'paradoxical undressing'. Climbers or hikers who are freezing to death have been known to take off their clothes just before they collapse. Not understanding this peculiar behavioural pattern, rescuers, and police, often conclude that a lone female hypothermia victim has been raped.

Clearly this sort of behaviour runs counter to survival, but so powerful is the brain's control over the body that at times it seems to be saying, 'My mind is made up. Don't confuse me with the facts.' There is no more blatant example than that which occurs when a hypnotized individual, touched with a pencil at room temperature, is told that it is actually a burning hot poker and subsequently develops a blister.

Thoughts have power. Intensive thinking can amplify, focus and direct that power enough to effect physical changes both within the body and without. Paralyzed people who are hooked up to amplifiers can be taught to concentrate their brain waves, creating enough electrical energy to purposefully move a cursor across a computer screen – a dramatic demonstration that willpower is real.

I saw even more convincing evidence in that blinding snowstorm below the summit of Mount Everest. I was sure Beck Weathers had died that day – and maybe he did. Maybe Bernie Chowdhury and Pablo Valencia also died. At the limit of their ordeals, they all had intense visions of being outside their bodies. This kind of sensation would arise if the circuits to the cerebral cortex were shut down in a last-ditch attempt to preserve energy for the basic body maintenance centres in the brain stem. The cortex would no longer receive any outside stimulation. Such internal quiet would lower the threshold of excitation for the nerves, setting off random bursts of impulses. These bursts of electricity can also be triggered by fear, fatigue, low oxygen and low blood sugar – all common conditions in survival situations. Some of the bursts might excite the temporal lobe, the portion of the cerebral cortex below the temples that controls conceptual thinking and language and image associations. Other impulses might excite the occipital lobe at the back of the head where visual images are registered. The effects could be seeing visions of loved ones or hearing the voice of God – phenomena that survivors often say gave them the strength to go on. With the cortex floating free, the survivor would also undergo that sense of detachment from the body that Beck, Bernie and Pablo Valencia all said they felt – the out-of-body experience often described by people who 'died and came back.'

Though I was certain that it was impossible for Beck to survive, Beck had other ideas. An idea is made from, or at least causes to form, a pulse of electricity in the brain. Fear of dying generates a lot of electricity in the amygdala, the emotion centre. Memories of home and family form electrical pulses in the frontal lobes that become more powerful as the images grow more intense. Nerve cells, or neurons, in every brain centre connect with cells in every other centre via axons, long wirelike offshoots that conduct electricity. Axons end in branches that don't quite touch any other nerve. They do, however, come very close to hundreds of dendrites, the receptor twigs of other neurons with which the axons nearly intertwine. When the electrical impulse reaches the end of the axon, it releases a chemical neurotransmitter that floats the signal across the tiny gap, or synapse, between axon and dendrites. The neurotransmitter has the potential to restart the electrical signal in each of the dendrites with which it comes in contact. Each of those receptor neurons can in turn start an impulse in hundreds of other neurons. The 100 billion neurons of the brain produce 100 trillion synaptic connections.

But a chain reaction will not lead to a coordinated response any more than an electrical signal travelling through a ball of uninsulated wires will turn on a machine. Not every thought leads to effective action. A micro-amp of thought that floats across a synapse won't react with every dendrite it washes up on. Whether or not a neurotransmitter will reconstitute an electrical signal in the next neuron depends on the reception it receives when it arrives. Memory, learning, experience and training will modify the response by having previously deposited chemicals that facilitate or inhibit the reconversion to electricity. Furthermore, the firing of the neuron is an all-or-nothing response. Once a threshold level of electrical current is reached, the signal is transmitted. Reaching that level might be achieved by one strong incoming signal, though it is often the additive result of many signals from many axons arriving on the same dendrite. A neuron receiving emotional input from the amygdala might be slightly below the threshold to fire. A small additional boost by a thought arriving from the frontal cortex, however, might put it over the top. The signal generated might induce a nonproductive response such as panic, or it might get channelled and focused to induce survival behaviour. One pulse of electricity, properly modulated, can create an electrical and chemical symphony. The modulating thought that provides enough energy to orchestrate that change is willpower. It can stir a body back to life.

Though Beck lay motionless in the snow, there was activity in his brain. Thought signals sparked his nerve cells, creating the random currents of a dreamlike state. Had Beck been hooked up to a PET scanner, it would have indicated energy flowing between the amygdala and frontal lobes as well as to the most primitive centres that maintain heartbeat and breathing. But his circuits were powering down and, with no new energy sources, would soon have shut off. The PET scan images would fade as Beck drifted into a peaceful pre-death unconsciousness.

Beck refused to die, however, and this created a change. Suddenly the weakening signals that were converging on the cingulate, the seat of the will, were amplified and redirected, becoming powerful enough to reinvigorate the parts of his brain that control motion and judgement, the functions he needed for survival. Beck got up out of the snow. He was able to think clearly enough to calculate in which direction to move. He headed off toward his own salvation.

Given that there was no change in his outside environment, where had Beck's burst of strength come from? How could the signal coming out of the cingulate suddenly be far greater than the sum of the signals that went in? What ignited the spark that pushed his nerves over the threshold? Perhaps it came from some undetectable source of energy within his nerve, or perhaps from a confluence of brain waves whose frequencies suddenly meshed to create an amplitude much greater than that provided by the individual components, much as a resonating sound can be far louder than the sum of the waves that form it. But still, how did Beck start it happening right then?

The laws of physics require a cause for every effect. Progressing backward, you eventually have to reach a first cause – some fundamental source of energy not dependent on anything before it, a primal force, not directly observable but perhaps the most natural and most fundamental of all. In the cosmos, the origin of all energy is called the 'singularity,' and the force it generates is called the 'big bang.' In the mind, the origin is the cingulate and the force is called will.

The true nature of will remains mysterious. Is it an electro-biochemical resource hidden deep within the brain or a force instilled from without by a higher power? The electrical spark generated by the cingulate may be the origin of will, or it may be the first detectable result of faith. The answer is within our bodies but beyond our grasp.

Every scientist knows that if you can't step outside a phenomenon to observe it, you can't hope to fully understand it. We need to use our brains to analyse our brains, so the information we take in will never be completely objective. Unable to escape from our own brains, there will always be an impenetrable, mystical barrier to understanding ourselves. The fundamental nature of the human will must remain unknowable. Ultimately, our explanations for surviving the extremes will require not just science, but faith.

NOTES AND BIBLIOGRAPHY

JUNGLE

Brainbridge, J.S., Jr- 'Frogs That Sweat.' *Smithsonian Magazine*, January 1989, 70–7.

Cambell, Jackson. 'Forest Medicine.'*Explorers Journal*, May (Summer) 1997, 18–25.

Eckholm, Erik. 'Secrets of the Rain Forest.' *The New York Times Magazine*, 17 January, 1988, 20–30.

Forsythe, Adrian and Miyata, Kenneth. *Tropical Nature*. New York: Charles Scribner's Sons, 1971.

Hooper, Joseph. 'The Gringo Chief.' *Outside*, August 1989, 35–40, 85–7.

Hughes, Carol and Hughes, David. 'Teeming Life of a Rain Forest.' *National Geographic*, January 1983, 49–65.

Jahoda, John C. and O'Hearn, Donna L. 'The Reluctant Amazon Basin.' *Environment*, October 1975, Vol. 17, No. 7, 16-20 and 25–30.

Kansil, Prince Joli. 'The Black Caiman of Zancudo Cocha.' *Explorers Journal* 66, no. 2 (1988): 50–3.

King, Steven R. 'Among the Secoyas.' *Nature Conservacy Magazine*, January–February 1991.

Krichner, John C. A. *Neotropical Companion*. Princeton: Princeton University Press, 1989.

Levy, Charles Kingsley. *Evolutionary Wars*. New York: W.H. Freeman, 1999.

Moffett, Mark W. 'Poison-Dart Frogs: Lured and Lethal.' *National Geographic*, May 1995, 98–111.

Peck, Robert McCraken. *Headhunters and Hummingbirds*. New York: Walker and Company, 1987.

Perry, Donald. *Life Above the Jungle Floor*. New York: Simon and Schuster, 1986.

Plotkin, Mark J. *Medicine Quest: In Search of Nature's Healing Secrets*. New York: Penguin, 2001.

———. *Tales of a Shaman's Apprentice*. New York: Penguin, 1994.
Raver, Anne. 'Adventures in the Amazon.' *Newsday*, 19 July 1987, 14.
Rennisi, Elizabeth. 'All Things Bright and Bitter.' *International Wildlife*, November–December 1989, 46–50.
Schreter, A. Harvey. 'The Jivaro (Shuar) Indians.' *Explorers Journal*, March 1986, 12–17.
Schultes, Richard Evan. 'Burning the Library of the Amazonia.' *The Sciences*, March–April 1994, 24–31.
Sterling, Tom. *The Amazon*. New York: Time Life Books, 1973.
Trupp, Fritz. *The Last Indians: South America's Cultural Heritage*. Austria: Perlinger Verlag, 1981.
Van Dyk, Jere. 'Rainforest.' *National Geographic*, February 1995, 2–39.
Wallace, Robert A. 'Searching for Medicines in the Vanishing Amazon.' *Explorers Journal*, Autumn 1993, 133–9.
White, Peter T. 'Nature's Dwindling Treasures.' *National Geographic*, January 1983, 2–48.

HIGH SEAS

Askenasy, Hans. *Cannibalism: From Sacrifice to Survival*. New York: Prometheus Books, 1994.
Bailey, Maurice and Bailey, Marilyn. *117 Days Adrift*. Dobbs Ferry, NY: Sheridan House, 1977.
Bombard, Alain. *Naufrage Volontaire*. Paris: Editions de Paris, 1958.
Callahan, Steven. *Adrift: Seventy-six Days Lost at Sea*. New York: Ballantine Books, 1986.
Center for the Study and Practice of Survival. *A Practical Guide to Lifeboat Survival*. Annapolis, Md: Naval Institute Press, 1997.
Cosquer, Henri. *La Grotte Cosquer*. Paris: Solar, 1992.
Craighead, F.C. *How to Survive on Land and Sea*. (4th ed.) Annapolis, Md: Naval Institute Press, 1984.
Duerenberg, P.M. Yap and van Stavern, W.A. 'Body Mass Index and Percent Body Fat: A Meta-Analysis among Different Ethnic Groups.' *International Journal of Obesity* 22 (1998): 1164–71.
Earle, Sylvia and Wolcott, Henry. 'Wild Ocean'. *National Geographic*, 1999.
García Márquez, Gabriel. *The Story of a Shipwrecked Sailor*. New York: Vintage International, 1989.
Garn, S.M. and Block, W.D. 'The Limited Nutritional Value of Cannibalism.' *American Anthropologist* 72, no. 106 (1970): 106–7.
Gill, Paul G., Jr, M.D. *Waterlover's Guide to Marine Medicine*. New York: Fireside, 1993.
Hanson, Neil. *Custom of the Sea*. New York: John Wiley and Sons, 2000.
Henderson, S. and Bostock, T. 'Coping Behavior after Shipwreck.' *British Journal of Psychiatry* 5 (1977): 543–63.

Heyerdahl, Thor. *Kon-Tiki* New York: Rand McNally, 1950.
———. *The Ra Expeditions*. Austria: Allen and Unwin, 1971.
Leslie, Edward E. *Desperate Journeys, Abandoned Souls: True Stories of Castaways and Other Survivors*. Boston: Houghton Mifflin, 1988.
Lum-McCunn, Ruthanne. *Sole Survivor*. New York: Scholastic, 1985.
McGarvey, Stephen. 'Obesity in Samoans and a Perspective on Its Etiology in Polynesians.' *American Journal of Clinical Nutrition* 53, no. 1 (1991): 1586S–94S.
Philbrick, Nathaniel. *In the Heart of the Sea: The Tragedy of the Whaleship* Essex. New York: Penguin, 2000.
Read, Piers Paul. *Alive! The Story of the Andes Survivors*. New York: Avon, 1975.
Robertson, Dougal. *Survive the Savage Sea*. Dobbs Ferry, NY: Sheridan House, 1994.
Ryan, Donald P. 'Kon-Tiki 50 Years Later.' *Explorers Journal*, Spring 1997, 11-21.
Stewart, George R. *Ordeal by Hunger: The Story of the Donner Party*. Boston: Houghton Mifflin, 1988.
Takada, Shiguro. *Contingency Cannibalism: Superhardcore Survivalism's Dirty Little Secret*. Boulder, CO: Paladin Press, 1999.

DESERT

Adolph, E.F. *Physiology of Man in the Desert*. New York: Interscience Publishers, 1947.
———. *Seasons of Sand*. New York: Simon and Schuster, 1993.
Bergstrom, J., Hermansen, L. Hultman, E. and Saltin, B. 'Diet, Muscle Glycogen and Physical Performance.' *Acta Physiologyca Scandinavica* 71 (October–November 1967): 140–50.
Broyles, Bill. *Journal of Arizona History* 23 (1982): 357.
———. *Journal of the Southwest* 30 (1988): 222.
Burgen, Michael. 'Life Is Cool in a Town Down Under.' *National Geographic World*, August 2000, 7–11.
Cabanac, M. 'Keeping a Cool Head.' *News in Physiological Science* 1 (1986): 41–4.
Case, R.M., and Waterhouse, J.M. *Human Physiology: Age, Stress and the Environment*. New York: Oxford University Press, 1994.
Clark, R.P. and Edholm, O.G. *Man and His Thermal Environment*. London: Edward Arnold, 1985.
Crawshaw, L.I. 'Temperature Regulation in Vertebrates.' *Annual Review of Physiology* 42 (1980): 473–91.
Johnson, Kirk. *To the Edge: A Man, Death Valley, and the Mystery of Endurance*. New York: Warner, 2001.
Kluger, M.J., Ringler, D.H. and Anver, M.R. 'Fever and Survival.' *Science* 188 (April 11, 1975): 166–8.

McArdle, W.D., Katch, F.I. and Katch, V.L. *Essentials of Exercise Physiology*. Philadelphia: Lee and Febiger, 1994.

McGee, W.J. 'Desert Thirst as Disease.' *Journal of the Southwest* 30 (1988).

Schmidt-Nielson, K., Schmidt-Nielsen, B., Jarum, B., Houpt, S.A. and Houpt, T.R. 'Body Temperature of the Camel and Its Relation to Water Economy.' *American Journal of Physiology* 188 (1957): 103–12.

Sides, Hampton. 'Crazy in the Desert.' *Men's Journal*, October 1999, 119.

Taylor, C.R. and Lyman, C.P. 'Heat Storage in Running Antelopes: Independence of Brain and Body Temperature.' *American Journal of Physiology* 222 (1972): 114–17.

Wolf, A. *Thirst*. Springfield, IL: Charles C. Thomas, 1958.

UNDERWATER

Becker, Pierre. *Tapping Submarine Freshwater Springs*. Aubagne, France: NympheaWater, October 2002.

Biax, Jeves. *The Conquest of Ocean Depth*. Marseilles: Comex, 1990.

Butler, P.J. and Jones, D.R. 'Physiology of Diving Birds and Mammals.' *Physiological Reviews* 77 (1997): 837–99.

Chowdhury, Bernie. *The Last Dive: A Father and Son's Fatal Descent into the Ocean's Depths*. New York: HarperCollins, 2000.

Falke K.J., Hill, R.D, Qvist, J., Schneider, R.C., Guppy, M. Liggins, G.C., Hochachka, P.W., Elliot, R.E. and Zapol, W.M. 'Seal Lungs Collapse During Free Diving: Evidence from Arterial Nitrogen Tensions.' *Science* 229 (1985): 556–8.

Gardette, Bernard, M.D. and Fructus, Xavier. *Hyperbaric Experiment Center: 36 Years of Deep Diving from Helium to Hydrogen*. Marseilles: Comex, 1991.

Hamilton, R.W., ed. *Development of Decompression Procedures for Depths in Excess of 400 Feet*. WS: 2–28–76. Bethesda, Md: Undersea Medical Society, 1976.

Hamilton, R.W., Heimbach, R.D. and Bove, A.A. 'Abnormal Pressures: Hyperbaric and Hypobaric.' In *Patty's Toxicology*, 5th ed., edited by E. Bingham, B. Cohrssen and C.H. Powell. New York: John Wiley and Sons, 2001.

Hindell, M.A., Burton, D.J., Bryden, H.R. and Bryden, M.M. 'Physiological Implications of Continuous, Prolonged and Deep Dives of the Southern Elephant Seal (Mirounga leonina).' *Canadian Journal of Zoology* 70 (1991): 370–9.

Hong, S.K. and Rahn, H. 'The Diving Women of Korea and Japan.' *Scientific American* 220 (1967): 88–95.

International Association of Free Divers: *Final Report on Audrey Mestre*.

Joiner, James T. *NOAA (National Oceanographic and Atmospheric Administration) Diving Manual*. 4th ed. Springfield, VA: NOAA, 2001.

Kamler, Kenneth, MD 'Balancing Risk and Reward.' *Immersed: The*

International Technical Diving Magazine 2, no. 4 (Winter 1997): 24–7.
Mayol, Jacques. *Homo Delphinus*. Florence, Italy: Glanet, 1986.
Rhan, H. *Physiology of Breath-Hold Diving and the Ama of Japan*. Pub. 1341. Washington, DC: National Academy of Sciences.
Rougerie, Jacques. *Expedition SeaOrbiter*. Paris: Peniche St. Paul, 2001.
Schmidt-Nielsen, K. *Animal Physiology: Adaptation and Environment*. 5th ed. Cambridge: Cambridge University Press, 1997.
Shilling, Charles W., Carlston, Catherine B. and Mathias, Rosemary A. *The Physician's Guide to Diving Medicine*. New York: Plenum Press, 1984.
Smith, E.B. 'On the Science of Deep-Sea Diving – Observations on the Respiration of Different Kinds of Air.' *Undersea Biomedical Research* 14 (1987): 347–69.
Smith, Gary. 'The Deadly Dive.' *Sports Illustrated*, 16 June 2003.
US Navy *Diving Manual: Air Diving*. Hagstaff, A2: Best Publishing, 1996.
Zaferes, Andrea. 'Rethinking the Hit.' *Immersed: International Technical Diving Magazine* 2, no. 1 (Spring 1997): 24–36.

HIGH ALTITUDE

Coburn, Broughton. *Mountain Without Mercy*. Washington, DC: National Geographic Society Books, 1997.
Hillary, Sir Edmund. *High Adventure*. New York: Oxford University Press, 2003.
Houston, Charles, MD. *Going Higher*. Seattle: The Mountaineers, 1998.
Hultgren, Herb, MD. *High Altitude Medicine*. Stanford, CA: Hultgren Publications, 1997.
Hunt, John. *The Conquest of Everest*. New York: Dutton, 1953.
Irving, L. 'Human Adaptation to Cold.' *Nature* 185 (1960): 572–4.
Jefferies, Margaret. *Mount Everest National Park*. Seattle: The Mountaineers, 1986.
Kamler, Kenneth M., MD. *Doctor on Everest: Emergency Medicine at the Top of the World*. London: Robinson, 2002.
Klesius, Michael. 'The Body: Adjust or Die.' *National Geographic*, May 2003, 30–3.
Krakauer, Jon. *Into Thin Air: A Personal Account of the Mount Everest Disaster*. New York: Anchor, 1998.
Monge, C. and Leon-Velarde, F. 'Physiological Adaptation to High Altitude: Oxygen Transport in Mammals and Birds.' *Physiological Reviews* 71 (October 1991): 1135–72.
Radomski, M.W. and Boutelier, C. 'Hormone Response of Normal and Intermittent Cold-Preadapted Humans to Continuous Cold.' *Journal of Applied Physiology: Respiration Environment Exercise Physiology* 53 (1982): 610–16.

Sherwood, Peter. *Everest: Legacies of Victors and Vanquished*. Hong Kong: FormAsia, 2003.
Siebke, H., Rod, T. and Lind, B. 'Survival after 40 Minutes Submersion without Cerebral Sequelae.' *The Lancet*, 7 June 1975, 1275–7.
Unsworth, Walt. *Everest*. London: HarperCollins, 1991.
Ward, M., Milledge, J.S. and West, J.B. *High Altitude Medicine and Physiology*. 2nd ed. London: Chapman and Hall, 1995.
Washburn, Brad. *On High: The Adventures of Legendary Mountaineer, Photographer, and Scientist Brad Washburn*. Washington, DC: National Geographic, 2002.
West, J.B. *High Life: A History of High Altitude Physiology and Medicine*. New York: Oxford University Press, 1998.
Wilkerson, James A., MD, ed. *Medicine: For Mountaineering and Other Wilderness Activities*. Seattle: The Mountaineers, 2001.

OUTER SPACE

Buckley, J.C., Jr. 'Preparing for Mars: The Physiological and Psychological Effects.' *European Journal of Medical Research* 4 (1999): 531–56.
Burrough, Bryan. *Dragonfly: NASA and the Crisis aboard Mir*. New York: HarperCollins, 1989.
Clark, Phillip. *The Soviet Manned Space Program: An Illustrated History of the Men, the Missions, and the Spacecraft*. New York: Orion Books, 1988.
Czeisler, Charles A. and Dijk, D.J. *Human Physiology and Sleepwake Regulation: Handbook on Behavioral Circadean Neurobiology*. New York: Plenum Publishing, 2001.
Fienberg, Richard Tresch. 'Endeavour's Excellent Adventure.' *Sky and Telescope*, April 1994, 24–7.
Hoffman, Jeffrey A. 'How We'll Fix the Hubble Telescope.' *Sky and Telescope*, November 1993, 23–9.
Hoffman, Stephen J., and Kaplan, David L., eds. *Human Exploration of Mars: The Reference Mission of the NASA Mars Exploration Study Team*. Houston, TX: Lyndon B. Johnson Space Center, 1997, 1998.
Jones, Eric M. 'Frustrations of Fra Mauro.' *Apollo Lunar Surface Journal*, 1995.
Long, Michael E. *The Body in Space*. Washington, DC: National Geographic, January 2001.
NASA. *Extreme Environment Mission Operations 5, 6*. Houston, TX: Lyndon B. Johnson Space Center, March 2003.
Nicogossian, Arnauld E., MD, Leach Huntoon, Carolyn, M.D. and Pool, Sean L., MD. *Space Physiology and Medicine*. 4th ed. Philadelphia: Williams and Wilkins, 2003.
Oberg, James E. *Red Star in Orbit: The Inside Story of Soviet Failures and Triumphs in Space*. New York: Random House, 1981.

O'Brien, K. and McLaughlin, J.E. 'The Radiation Doses to Man from Galactic Cosmic Rays.' *Health Physics* 22 (1972): 225–32.
Polk, Milbry C. 'A Brave New World.' *Explorers Journal*, Winter 2002, 10–12.
Remmers, J.E., ed. 'Applied Physiology in Space.' *Journal of Applied Physiology*, Volume 8 (1), July 1996.
Satava, Richard. *The BioIntelligence Age: Medicine after the Information Age*. Amsterdam: IOS Press (70), 2000–7.
———. 'Emerging Technologies for Surgery in the 21st Century.' *Archives of Surgery* 134 (November 1999): 1197–1202.
White, Ronald J. 'Weightlessness and the Human Body.' *Scientific American*, September 1998.
www.nasa.gov
www.themarssociety.org
Zorpette, Glenn. 'A Taste of Weightlessness.' *Scientific American*, November 1999.

NEUROBIOLOGY

Begley, Sharon. 'Religion and the Brain.' *Newsweek*, 7 May 2001, 50–7.
———. 'Thinking Will Make It So.' *Newsweek*, 5 April 1999, 64.
Cainsworth, A.G. *Secrets of the Mind*. Copernicus, NY: Springer Verlag, 1999.
Carter, Rita. *Mapping the Mind*. Berkeley: University of California Press, 1998.
Czerner, Thomas, MD. *What Makes You Tick*? New York: John Wiley and Sons, 2001.
Damasio, Antonio. *Descartes' Error*. New York: Avon Books, 1994.
Dawkins, Richard. *The Selfish Gene*. New York: Oxford University Press, 1990.
Evars, Stephen. 'The Cerebral Hemodynamics of Music Perception.' *Brain* 122 (January 1999):1.
Kamler, Kenneth, MD, Koenig, Harold, MD and Pressman, Peter, MD *The Group Room: The Role of Faith, Belief, Prayer, and Hope in the Face of Cancer or Any Life Threatening Situation*, May 4, 2003.
Koenig, Harold, MD. *The Link Between Religion and Health: Psychoneuroimmunology and the Faith Factor*. New York: Oxford University Press, 2002.
LeDoux, Joseph. *The Emotional Brain*. New York: Simon and Schuster, 1996.
Nunan, David. 'The Ultimate Remote Control.' *Newsweek*, 25 June 2001, 71–5.
Pert, Candace B. *Molecules of Emotion*. New York: Simon and Schuster, 1997.
Pressman, Peter, MD. 'Religion, Anxiety, and Fear of Death.' *Religion*

and Mental Health. Ed. John F. Schumaker. New York: Oxford University Press, 1992.
Ramachandran, V.S., MD, and Blakeslee, Sandra. *Phantoms in the Brain*. New York: William Morrow, 1998.
Wade, Nicholas, ed. *The New York Times Book of the Brain*. The New York Times, 1998.

GENERAL

Ascroft, Frances M. *Life at the Extremes: The Science of Survival*. Berkeley: University of California Press, 2000.
Auerbach, Paul S., ed. *Wilderness Medicine*. St. Louis: Mosby, 2001.
Bennet, Glin. *Beyond Endurance: Survival at the Extremes*. New York: St. Martins Press, 1983.
Boorstin, D. *The Discoverers*. New York: Penguin Books, 1983.
Burnham, Terry and Phelan, Jay. *Mean Genes: From Sex to Money to Food: Taming Our Primal Instincts*. New York: Penguin, 2001.
Case, R.M., and Waterhouse, J.M. *Human Physiology: Age, Stress and the Environment*. New York: Oxford University Press, 1994.
Department of the Army. US *Army Survival Manual*. New York: Dorset Press, October 2001.
Forgey, William. *Wilderness Medicine: Beyond First Aid*. Guilford, Conn.: Globe-Pequot Press, 2003.
Gross, M. *Life on the Edge*. New York: Plenum Press, 1998.
James, William. 'The Energies of Men.' *The American Magazine*, 1902.
———. *The Varieties of Religious Experience*. London: McMilliam Publishing, 1961.
Maniguet, Xavier. *Survival*. New York: Barnes and Noble Books, 1999.
Schmidt-Nielsen, K. *Animal Physiology: Adaptation and Environment*. 5th ed. Cambridge: Cambridge University Press, 1997.
Solzhenitsyn, Aleksandr. *The First Circle*. Evanston, IL: Northwestern University Press, 1997.
Stark, Peter. *Last Breath: Cautionary Tales from the Limits of Human Endurance*. New York: Ballantine, 2001.
Stilwell, Alexander. *The Encyclopedia of Survival Techniques*. New York: The Lyons Press, 2000.
Stroud, M. *The Survival of the Fittest*. London: Jonathan Cape, 1998.
Troebst, Cord Christian. *The Art of Survival*. New York: Doubleday, 1965.
Wharton, David. *Life at the Limits: Organisms in Extreme Environments*. Cambridge: Cambridge University Press, 2002.
Widmaier, Eric P. *Why Geese Don't Get Obese*. New York: W.H. Freeman, 1999.
Wright, Robert. *The Moral Animal: Evolutionary Psychology and Everyday life?* New York: Vintage, 1995.
www.explorers.org

ACKNOWLEDGEMENTS

THE SPIRIT OF EXPLORATION is a powerful force that links individuals all over the world. It can override geography, ethnicity, culture, religion, politics, social standing and any and all other civilized differences. The common bonds that it creates form an invisible network that covers the earth, and being one strand in that network has given me access to the most remote corners of the planet, the people who study those corners and the people who live there. I have benefited enormously from observing nature up close – through my own eyes but also, and maybe even more so, through the eyes of other explorers – from those who have MD or PhD educations to those who don't know how to read. My teachers have been doctors and scientists trained in universities and research labs, field explorers taught by experience and necessity and local natives imbued with the wisdom of their elders. The synthesis of knowledge that this book represents is the fruit of their teachings, of their enthusiasm for the natural world and of the spirit they show in surviving the extremes.

The idea for the book grew out of my dual enthusiasms for exploration and medicine. In the course of casual conversations and formal lectures, I realized that many people are as intrigued by the workings of the human body as they are by the remote, mysterious corners of the world. It would provide fascinating reading, I thought, to take a continuous journey from the outer limits of the earth to the inner limits of the body. I held the thought for a while, until at an Explorers Club dinner, I was not so casually asked by the person seated next to me if I had any ideas for a book. He was the literary agent Andrew Stuart, and he was immediately taken by the book I envisaged. He energized my idea, supplying continuous sparks to keep me writing and trenchant analysis to channel my efforts. Together we developed the framework of the book and its unifying theme. Once he was satisfied with the manuscript, he became my guide through the extreme environment of publishing, providing me with direction, encouragement and reassurance as needed. He had an uncanny understanding of what editors and publishers were looking for, and I was repeatedly amazed by his prescient advice. All his optimistic predictions came to pass. I've learned that if Andrew tells you something, believe it.

To gather the additional information I would need to supplement my own files, I brought Alison Mitchko on board – a research assistant with a deft hand at the computer and a wry sense of humour. She was often tested to the limit on both counts – most especially on one occasion when I woke her up in the middle of the night asking for some arcane information on penguin anatomy. I insisted it be faxed to me immediately since I was calling from a bakery six time zones away that was about to close. She got it done, remembering that our research had shown that maintaining a sense of humour is critical to survival in extreme situations.

My manuscript was dictated and submitted in sections for typing, always on short notice and behind schedule, but always completed with amazing speed and accuracy by my typist, Rochelle Garmise.

Three editors were on my team. Sam Weiss, my longtime friend and neighbour and an editor for the *New York Times*, volunteered to read my manuscript and improved it with a newspaperman's objective analysis and timely suggestions. My freelance editor and background researcher was David Groff. He and I communicated on the same frequency throughout the preparation of the manuscript. He had an intuitive feel for my writing and read the text with sensitivity and uncommon insight. His subtle suggestions and grand ideas were always on target and the book is much the better for them. At St. Martin's Press, senior editor Tim Bent believed strongly in the book right from the beginning. Though I was an instinctive writer with no formal training in literature and he was a self-described 'reformed academic,' he had the insight to see that we could develop a synergy that would elevate the book into the realm of literature. To the extent that we have succeeded, I owe him my thanks. But even more, I am grateful to Tim for showing me that, in developing and polishing a manuscript, everything from defining the theme to appreciating the power of semicolons can be a joyful experience. We educated each other, and our interchange of ideas stimulated and uplifted us throughout the long writing and editing process. Our marathon sessions sometimes left me exhausted, but always left me exhilarated.

I saw this book as an opportunity to transmit to others my sense of awe and wonder for the forces that drive the human body. The more we understand the body the more fascinating it becomes. Therefore, I endeavoured to provide as much detail as possible without losing clarity or overall understanding. Distilling huge volumes of medical information into straightforward logical descriptions and explanations runs the risk of inadvertently being misleading or inaccurate. To guard against this possibility, the text was graciously reviewed by a number of doctors and scientists – established experts in their respective fields. At my invitation, each of them undertook the task not only willingly but eagerly and enthusiastically. In alphabetical order, my expert reviewers were: Michael V. Callahan, MD, Professor of Infectious Diseases-Biothreat Detection, Massachusetts General Hospital, Boston, Massachusetts. William Forgey, MD, past President, Wilderness Medical Society, Grown Point, Indiana. John J. Freiberger, MD, Professor, Centre for Hyperbaric Medicine and Environmental Physiology, Divers Alert Network, Duke University Medical Centre,

Durham, North Carolina. Bernard Gardette, MD, Medical Director and Chief of Hyperbaric Medicine, Comex, Marseilles, France. Davidson Hamer, MD, Director, Travellers Health Service, New England Medical Centre, Boston, Massachusetts. R.W. (Bill) Hamilton, PhD, Director, National Board of Diving and Hyperbaric Medical Technology, Tarrytown, New York. Michael Hawley, PhD, Director of Special Projects, Massachusetts Institute of Technology, Cambridge, Massachusetts. Jeffrey Hoffman, PhD, Astronaut, Professor of Aeronautics and Astronautics, Massachusetts Institute of Technology, Cambridge, Massachusetts. Charles Houston, MD, Emeritus Professor of Medicine, University of Vermont, Burlington, Vermont. David Johnson, MD, President and Medical Director, Wilderness Medical Associates, Bryant Pond, Maine. Harold Koenig, MD, Professor of Psychiatry and Medicine, Duke University Medical Centre, Durham, North Carolina. Michael J. Manyak, MD, Professor of Microbiology and Tropical Medicine, George Washington University, Washington, DC. Victoria McKiernan, Scuba Divemaster, Washington, DC. Christian F. Ockenhouse, MD, Lt. Colonel US Army, Department of Immunology, Walter Reed Army Hospital, Silver Springs, Maryland. Peter Pressman, MD, Professor of Medicine, University of Southern California, Los Angeles, California. Richard Satava, MD, Retired Lt. Colonel US Army, Project Director, Defence Advanced Research Projects Agency, Professor of Surgery, University of Washington, Seattle, Washington. William Todd, PhD, Director of Spaceflight Simulations, Johnson Space Centre, Houston, Texas. Richard Williams, MD, Medical Director, NASA Astronaut Program, Washington, DC. I am indebted to these experts for their careful analysis of my work. Their comments, corrections and suggestions have added greatly to the accuracy of this book. Any errors, however, remain my sole responsibility.

To range so widely through geography and science I needed, and received, the help of a remarkably diverse collection of people, some listed above, whose assistance I specifically asked for, and some listed below, who might be surprised to see their names mentioned – perhaps unaware of the contributions they made to my education or experience. The logical place for me to begin thanking people is at the Explorers Club. I am forever grateful to a succession of Club presidents who fostered and encouraged my interest in exploration: Charles Brush, an anthropologist-archaeologist, who introduced me to the Club and invited me to join; John Levinson, a ship's surgeon and consultant on high seas medicine, who encouraged me in my idea to collect the members' medical experiences; John Loret, a professor of marine sciences and Director of the Science Museum of Long Island, who coincidentally had taught me to scuba dive while I was in college and who transmitted to me then his enthusiasm for exploration and for the underwater world; and Alfred McLaren, a nuclear submarine captain and polar explorer, who put me in charge of research and education at the Club, an ideal position in which to meet, and learn from, the people who are actually out there at the frontiers.

I have learned much from Sylvia Earle, Explorer-in-Residence at the National Geographic Society and former Chief Scientist for the National Oceanographic and Atmospheric Administration. Her knowledge of seas is as vast as the seas themselves and I was constantly buoyed up by her enthusiasm for my project. I had many conversations with Jim Fowler, cohost of the TV show 'Wild Kingdom' and Director of the Wildlife Conservation Centre. He has given me great insight into the jungle and the interrelationship of man and the natural world. With Donn Haglund, Chairman of the Department of Geography at the University of Wisconsin, I studied the ways of the people and animals of the Arctic. With Max Galimore, airline pilot and scuba diver, I surveyed life in the sea and on land in the Galapagos Islands. Jeff Hoffman, an astronomer and one of the NASA astronauts who fixed the Hubble telescope, provided me with a personal account of life in outer space.

The rewards of exploration are derived not just from where you go but in large measure from whom you go with. I have had the rare good fortune to endure harsh environments with some of the finest people on Earth, and have been strengthened by their example. Expeditions are agglomerations of people from disparate backgrounds, and I am proud of the eclectic group of friends I have accumulated over the years.

In the jungle, I met Antonio, for whom survival was a daily event. I was brought to him by Sebastián, my native Ecuadoran guide, who bridged two worlds and so could help me better understand the people who were around us. The expedition was led by Bill Jahoda, Professor of Biology at Eastern Connecticut State University, and his son, John Jahoda, Professor of Biological Sciences at Bridgewater State College, Massachusetts. I am indebted to them, and to all the other zoologists and botanists who patiently explained their research to me and made sure that I wasn't just looking at the jungle, that I was actually seeing it.

I owe my French connection with the undersea world to my good friend Louis Potie, a hydrologist and former Deputy Mayor of Marseilles, who introduced me to the pioneers of the deep. Of the hundreds of dives I have made, those that helped the most in the preparation of this book were done in the Mediterranean Sea, with the cooperation of Comex and its founder and chairman, Henri Delauze. Comex's Chief Medical Officer, Bernard Gardette, is an extreme medicine researcher who brought me, academically, to the edge of the ocean frontier. I am especially grateful to Pierre Becker, President of NympheaWater, for making me part of his dive team exploring the resources of the seafloor. And a thank-you goes to marine architect Jacques Rougerie for his invitation, which I have accepted, to be part of the *SeaOrbiter* crew when it starts its around the world drift.

From the depths to the heights. My first climbing instructor, ex-Green Beret Rob Blathewick, saw fit to include an inexperienced doctor on what became my first expedition. Since then I have climbed all over the world and tested the limits of extreme medicine on six expeditions to Mount Everest. The first four were led by Todd Burleson of Alpine Ascents International, with Pete Athans as the climbing leader. They were sponsored in part by the National Geographic Society, under the direction of Brad Washburn, their former Chief Cartographer and President Emeritus of the Boston Museum of Science. I have worked closely with Brad for many years. He is a former world-class mountain climber and has come to represent for me and for many others the ultimate example of an explorer-scientist.

My latest two expeditions to Everest were undertaken at the request of the NASA Commercial Space Centre under the direction of Dr Ron Merrill at Yale University. The expeditions were led by Scott Hamilton, a director of the Explorers Club, and were specifically designed to test space-age equipment and communications at the limits of human survival. Via satellite, we were supported at Yale by Dr Richard Satava, also a member of the Defence Advanced Research Projects Agency, and by Dr Peter Angood. On Everest, where I was the chief high-altitude physician, I worked with chief medical researchers Dr Vincent Grasso, a fellow at Yale, and Dr Chris Macedonia, a US Army major on loan from the Defence Department. The engineering work was carried out by Mike Hawley and his group of techno-wizards from the Massachusetts Institute of Technology Media Lab.

Everest is a difficult and dangerous place to work. A large support team of American and Sherpa climbers successfully managed to expedite our research and keep us safe at the same time. I must especially recognize the conscientious and sometimes heroic efforts of Wally Berg and Jim Williams, two elite mountain guides who are able to maintain strength and clarity of thought even at the most extreme altitudes. Most especially, however, I want to thank my Sherpa climbing friend Nima Tashi, who with quiet loyalty and devoted support has looked after me for many years on the mountain.

To write a book on survival you have to leave extreme environments safely. Not everyone does. I wish to pay tribute to some friends who lost their lives in the course of their expeditions. New Zealand Everest climbers Gary Ball, who succumbed to pulmonary oedema on Dhalighiri, and Rob Hall, who became hypothermic on Everest; American Muggs Stump with whom I climbed in Antarctica, who fell into a crevasse on Mount McKinley; my British pilot in Antarctica, Giles Kershaw, who crashed on the ice when his plane was caught in a downdraught; and Argentinian Explorers Club member Adrian Hutton, who crashed while flying low over a mountainous site in the Andes where he was reconnoitring his next exploration.

I have been lucky enough to come back from every expedition I have been on, lucky to have loyal friends eager to see me and luckier still to have a devoted family waiting for me when I come home: my mother, Ethel; my brother, Jerry, and my sister-in-law, Marilyn; my two children, Jonathan and Jennifer; and my father, Willie, watching over us all. They are with me on every adventure, and they have sustained me through difficult times away and at home. They have endured my absences for weeks or months at a time as well as the countless hours I have spent writing. Those were hours my children and I shared in our study, I on the couch writing, they at their desks doing homework – the three of us exchanging ideas, opinions, advice and laughter. The memory of those days together I will keep as a timeless treasure. I hope that we will always share all our adventures and take pride in each other's accomplishments. My family, and my many friends, give me the strength I need to explore extreme environments, and, especially, they give me that most valuable key to survival – the motivation to get back.

INDEX

Aborigines 222
acclimatization
  to mountains 213, 214, 217, 233
  to outer space 279
acetone 122
achiote 42, 59, 70–1
acupressure 101
Adams, John 133
adapting to the environment 15, 55, 57, 70, 302
  cold tolerance 221–3, 235, 236
  deserts 137, 138, 151, 155–6, 157, 160, 161, 162, 163, 164–5
  jungle 52, 56, 88–9
  mountains 231–2, 233, 234
  oceans 132
  outer space 285, 293
  underwater 197–8
adrenaline 153
agouti 67
air, compressed 186, 193
air pressure 172, 177, 178, 193, 211
Albatross 92–3, 97, 98, 99, 133
Algeria 150
alkaloids 75, 76
altruism 65, 249, 250, 307–8
alveoli 212, 225–6, 227, 267
Amazon 19–90
ampiwaska vine 73–4
amputation 119–20, 301–2
anacondas 58–9
anaerobic respiration 232
anaesthesia 30, 98
anatomy 124
Anderson, Clay 261–2, 263–4
Andes 124
Annapurna 7
Antarctica 291, 292, 313–14
antelope 152–3, 310
anti-G suit 268, 270
antibodies 51, 52
antishock trousers 268
antivenins 51–2
Antonio (Amazonian Indian) 21, 23–4, 25, 26, 27, 31, 40, 42, 46, 47–8, 52, 55, 57, 58–61, 62, 64, 66–8, 71, 73–6, 78–84, 86, 87–8, 190, 302, 334
ants 62–4, 87, 159
aorta 285
Apollo 13 282–4
Aquarius 260, 261, 262–3
aqueous and vitreous humours 116
Arab–Israeli war 150–1
arctic foxes 223
army ants 62–3, 87
arteries 141, 240
arterioles 141
astronauts 260, 266–300
Athans, Pete 244, 245, 246, 247, 248, 250, 335
Atlantic Ocean 99
aurora borealis 286–7
auto-recovery system 269
autonomic nervous system 315
axons 318
ayahuasca 75, 76, 79
Ayamara Indians 232–3

Bailey, Marilyn and Maurice 110, 114, 128, 129, 131, 132, 133, 302, 307
Baker, Norman 119, 120
barnacles 104–5, 109
basal metabolic rate 122, 139
bears 310
Becker, Pierre 183–4, 193, 334
beetles 157
bends 15, 187–91, 277, 280
Bernabe, Pascal 182
Berullio (Amazonian Indian) 21, 22–3, 24, 25, 27, 29, 30, 31, 36, 45, 57
betadine 32
bicarbonate 213
biofeedback 316
blisters 53
blood
  arterial blood 212
  circulation 141
  coagulation 50
  cooling 141
  glucose 121
  red blood cell concentration 240, 292
  salt concentration 117–18, 126, 130, 143, 144, 146
  thickened 143, 144, 146, 148
blowguns and darts 66, 72, 75, 80–1, 82
blubber 180
body colour 164–5
body temperature 138, 139–40, 142, 143, 155, 222, 315
  thermoregulation 140–1, 147, 165, 179, 265
Bombard, Alain 120–1
bone marrow 213–14
bones 291–2, 297
borderline environments 5–6, 14
boredom 289–90, 294
botflies 304
Boukreev, Anatoli 245, 248–9, 302
brain 3, 6, 55, 101, 118, 119, 132, 224, 273, 308–11
  amygdala 77, 129, 309, 312, 313, 315, 316, 318, 319
  brain stem 309
  cerebellum 238, 273
  cerebral cortex 93, 125, 129, 143, 169, 237, 238, 308, 309–10, 317, 318
  cerebral metabolism 219
  cerebral oedema 237–8, 239, 246
  cingulate 311, 312, 314, 315, 316
  complexity 77, 308–9
  cooling 152
  electrochemistry 120, 129
  frontal cortex 77, 78, 95–6, 97, 310, 319
  frontal lobes 221, 273, 310–11, 313, 314, 315, 318, 319
  hippocampus 77, 129
  hypothalamus 95, 97, 140–1, 142, 143, 145, 147, 152, 156, 204, 212, 213, 226, 235, 285
  medulla 238
  neurophysiology 77–8, 308–11, 314–16
  occipital lobe 318
  oxygen deprivation 237
  oxygen supply 93, 97
  parietal lobes 78
  temporal lobes 78, 318
  visual cortex 120
brain cell death 96, 98
Breashears, Dave 250, 254
buoyancy-compensation vest 197
Burleson, Todd 220, 244, 246, 247, 248, 250, 335
bushmaster 52

caissons 185
caisson's disease 187
calcium 292
Callahan, Steve 91–2, 95–6, 98, 99, 102–3, 104, 113, 128, 129, 131, 149, 302
callanques 173–5
calluses 53
camels 155, 157, 158, 160, 161, 164, 222
canawaska vine 66
cancer 287
candiru 45, 86
cannibalism 123–7, 313
carbohydrates 111, 112, 122, 123, 124, 131
carbon dioxide 95, 213, 283
caribou 223

carotid arteries 212, 267, 285
Cassis 173
catecholamines 153
caterpillars 54
catfish 45, 84
cave paintings 175
centrifugal force 273–4
centrifuge 273
cerebral oedema 237–8, 239, 246
cerebrospinal fluid 116, 238
chanting, Tibetan 1, 4–5, 15–16
chaperone proteins 153, 154
chicha 69
chloroquine 39
chonta palms 79
Chowdhury, Bernie 188–91, 302, 307, 317, 318
clothing
  astronauts 278–9, 280
  desert peoples 160
  mountaineering 202–3
Cofan Indians 21, 40
cold tolerance 221–3, 235, 236
cold-water survival 179–80
Comex 193, 194
conditioning 302
  see also adapting to the environment
conga ants 62
Coober Pedy 163
coral snake 45–7, 49
cormorants 104
cortisol 153, 294
cosmic rays 265, 286, 287, 288
Cosquer, Henri 173–5
Coumadin 74
coumarin 74
crocodiles 21, 22, 23, 24, 25, 37–8, 39–40, 43, 58
curare 72, 74, 81

dates 162–3
decompression 188, 189, 191–2, 195, 197, 260, 276, 279
dehydration 100, 118–19, 143, 144, 213
Delauze, Henri 193, 334
delusions 316–17
demand regulators 176
dendrites 318
depressurization 276, 277–8, 280
desert survival 135–70
  body temperature regulation 140–1, 146, 147
  food collection 158–9, 162–3
  physiological survival strategy 144, 154, 155, 156
  water gathering 143, 157–8, 161–2
  water loss 142, 143, 146, 151
dexamethasone 239, 246, 250
Dexedrine 275
diarrhoea 41, 73
diet, human 111, 113
diurnal rhythms 290–1
diving 171–200
  bends 15, 187–91
  breath-holding diving (free diving) 175–6, 180, 181–3
  commercial deep-sea divers 196
  deep-sea diving 15
  hard-hat diving 176
  mammalian dive reflex 179, 181
  saturation divers 195–7, 200, 260
  scuba diving 96, 176, 260
diving bells 176, 195–6
DNA 287
dolphins 180
Donner Party 125, 132, 313
Dramamine 101
drowning 93, 94, 95, 97–8
dysentery 40

ears 177–8
  ear pain 178
  eustachian tube 178
  middle ear 177–8
  popping 172–3
  vestibular system 271, 272, 273, 274, 285
Ecuador 20, 90
  see also pulmonary oedema
electric eels 44, 73
electrochemistry 120, 129
electrolytes 100
emotions 312, 313, 314, 315
emphysema 210
endolethial cells 252
endurance athletes 135, 154, 155
erythropoietin 213, 214
Eskimos 104, 164, 223, 224
Etienne, Jean-Louis 222

evolution 313
Explorers Club 10–11, 333
extreme medicine 6–7, 11, 12, 16, 17, 99

face (diving) masks 178, 198
fats 112, 114, 122, 123, 131, 132, 164, 223
fear 93, 128, 129, 154, 318
feet
  cold water exposure 235
  insulation 203
feral state, reverting to 127
Ferreras, Francisco 180–1
fibrin 252
Fischer, Scott 243, 245, 247, 249
fish
  bioluminescence 197, 198
  deep-sea fish 198–9
  swim bladders 197, 199
fishing 103–4, 105, 106–7, 109–10, 111, 113
flatus 234–5
flying fish 114, 123
Foale, Michael 275, 277
foot fungus 53
frogs 64–5, 66, 74, 80, 81, 89, 223, 308
frostbite 245, 246, 248, 251–4, 255

G force 266, 267–8, 269, 270, 272, 274
gag reflex 94, 96, 98
Galapagos Islands 108, 178
gangrene 301
Gardette, Bernard 193, 194, 334
Gau, Makalu 243, 247, 250, 251–4, 257, 258
gene pool 70
glucagon 153
glucose 121, 153, 154
glycogen 121, 154
goose bumps 219
gravity 265–6, 268, 269, 271, 272, 273, 274, 297, 298, 301
Gulf Stream 199
gumbolimbo tree 73, 86

haematoma 3, 4–5
haemoglobin 179, 214, 232
Haise, Fred 282
Hall, Rob 243, 244, 246, 247, 249, 250, 307, 336
hallucinations 119, 186, 238
Hansen, Doug 243, 244, 246, 247, 249
Hansen, Henrik 251, 253
heads, shrunken 82–3
heart attacks 240
hearts 231–2, 233
heat
  conduction 138, 139, 155
  convection 138, 139
  radiation 138–9, 139, 155, 265, 286, 287–8
  see also temperature
heat exhaustion 144, 146, 205
heat stroke 146
heat-shock proteins 153, 155
helicopter evacuation 258
heliox 193, 194, 195
helium 193–4
Hermanigildo (Amazon Indian) 26, 27–36, 40, 57, 69, 70–1, 263–4, 302, 303, 315
Heyerdahl, Thor 110, 118
high-pressure nervous syndrome 194
Himalayas 209
Hoffman, Jeff 280
hook manoeuvre 268
hormones 132–3
hospitals and nursing homes 307
Hottentots 164
Hubble Space Telescope 280, 287
Hughes family 137–8, 145–6, 147, 307
human body 5, 14
  body shape remodelling 164, 223
  physiological constraints 5
  physiological responses 14–15
  water content 142
  see also body temperature
human society 63
hunger 121, 129
hunter reflex 235
Hwang, Emma 261–2, 263
Hyakutake 242
hyaluronidase 50
hydrogen 194, 288
hyperbaric chambers 195, 196, 200, 239

hyperthermophiles 155
hyperventilation 93
hypnosis 317
hypothermia 247–8, 258, 317

igloos 224
immune system 292
International Space Station (ISS) 261, 275
IV fluid 216, 228, 251, 257

Jahoda, Bill 29, 32, 35–6, 44, 45, 334
jerboas 163
jungle *see* Amazon

Kalahari bushmen 162, 222
Kami (Sherpa) 240–1
Kennedy, John F., Jr. 273
Khumbu cough 224
Khumbu Glacier 211
kidney stones 292
kidneys 118, 130, 144, 147, 152, 213, 267, 285
Kipling, Rudyard 100
knowledge as survival tool 302
Koncha (kitchen boy) 225, 227–8, 229–30, 231, 234
krill 110
Kufka, Julianne 303–4

lactic acid 232
lactic acid paradox 232, 233
latrines 54, 55, 85
Lazarev, Vasily 270
Lazutkin, Aleksandr 275, 277
Leonov, Aleksei 279
lidocaine 30, 31, 264
life rafts 91, 92, 96–7, 102–3
Lim, Poon 110–11, 133
lithium hydroxide 283
llamas 233
Lovell, Jim 282
lucid interval 96
luck 303
lungs 94, 95, 179, 181, 212, 225–6, 226, 231, 232, 233, 268
  pulmonary oedema 6, 225, 226–8, 230–1, 232, 234, 246

McGee, W.J. 166, 167, 168, 169
machete wound 26–37
magnetic resonance imaging (MRI) 310
Makarov, Oleg 270
malaria 38, 39
manioc 69, 81, 82
Marathon des Sables 135
Mars 259, 296–300, 301
Mary-Jeanne 121, 123
Masai 164
Massachusetts Institute of Technology 17
Mayol, Jacques 180, 181
medicinal plants 73–6, 79
melanin 164–5
melatonin 281, 291
menstrual cramps 73
Mestre, Audrey 181–3
metabolic efficiency 132
Mignonette 127, 128, 313
migration 224
military commandos 304–5
Mir space station 275, 276–7
mirages 149
mitochondria 214
Mojave Desert 165–70
mosquitoes 38–9, 82, 303–4
Mount Aucanquilcha 233–4
Mount Everest 1–3, 13, 16, 17, 201–58, 278, 335
mountain survival
  frostbite 245, 246, 248, 251–4, 255
  heat conservation 202–3, 219
  hypothermia 247–8, 258, 317
  mountain sickness 210
  see also Mount Everest
muscles 122, 124, 130, 132, 214
  atrophy 291
  contractions 49, 223
  convulsions 147
  fast-twitch muscles 154
  slow-twitch muscles 154
myoglobin 179, 214–15

Namba, Yasuko 244
NASA 17, 260, 284, 335
National Geographic 13, 242
National Oceanographic and Atmospheric Administration (NOAA) 260
natural selection 14, 55–6, 105, 165, 231, 305

navigation 131, 132, 159
Nazi experiments 150
NEEMO–NASA Extreme Environment Mission Operations 260–2
Nepal 208–9
nerve repairs 34–5
nerves 34
neuroma 34, 35
neurons 318, 319
neurotransmitters 49, 65, 75–6, 77, 318, 319
neutron bombs 288
nifedipine 228, 229, 253
nitric oxide 232, 233
nitrogen 15, 113, 186, 193, 194, 260, 276
nitrogen narcosis 185, 186, 187, 189, 192, 193
nomads 160, 161, 162, 164
Nymphea Water 183, 184, 186

obesity 132
ocean currents 199
ocean survival 91–134
  food gathering 103–5, 106–7, 109–10, 111, 113–14
  physical problems 99–101, 109
  psychological problems 109, 129
  voluntary castaways 120–1
  water gathering 101–2, 103, 106, 111, 114, 115, 116–17, 120
  see also diving
oedema 130
offshore oil rigs 194
opal miners 163
osmosis 117, 144
osteoporosis 291–2, 297
out-of-body experiences 169, 190, 256, 317, 318
outer space 259–300
oxygen 95, 97, 214
  breathing pure oxygen 193, 227
  deficit 33, 93, 213, 266–7
  diffusion 212
  oxygen-haemoglobin link 214, 232
oxygen cylinders 216–17

paca 80, 81, 82
Pacific Ocean 99
pain 30, 31, 128, 314
  phantom limb pain 119–20
  suppression 57
paradoxical undressing 317
paralysis 317
Pasang (Sherpa porter) 1–2, 3, 4, 13, 15–16
pearl divers 175–6, 179, 214
peroxide 32
Peru 7–10
phantom limb pain 119–20
Phenergan 274
Pheriche 229
piranhas 43, 87
pirarucu 84
pit vipers 47–8, 50
placebo effect 101
plankton 110, 114
Plessz, Catherine 92, 96–7, 98, 99
pneumonia 301
poaching 23
poison-dart frogs 64–5, 66, 74, 80, 81, 89, 308
polio 53
Polynesians 110, 131, 132, 164
polypropylene clothing 202, 203
positive thinking 70
positive-pressure breathing 268, 270
positron emission tomography (PET) scans 310, 319
prayer and meditation 78
pressure cookers 265
pressure drops 277
pressure sores 109
Prosperi, Mauro 135, 136–7, 147–51, 151, 154–5, 156, 157–60, 165, 167, 170, 302
proteins 112–13, 122, 123, 124, 131, 132, 139
pulmonary oedema 6, 225, 226–8, 230–1, 232, 234, 246

Quechua Indians 232–3

Ra II 118, 119
radial artery 33, 35
radial keratotomy 255
radial nerve 33
radiation 139, 265, 286, 287–8
Rais 231
Ralston, Aron 305–6

reasoning 128, 312, 313
rehydration enemas 115
Reisman, Garrett 264
religious experience 78
remora 110
respirators 269
rheumatism 73
Ricardo (South American climber) 237–40
rifles 74–5
Ríos, Jesus 166–7, 168, 169
ritual participants 76, 77
Robertson, Dougal and Lynn 105–7, 114, 115–16, 128, 129, 131, 302, 307
robotic surgeons 295–6
Rougerie, Jacques 200, 334
Rutten, Otto 262, 263, 264

Sahara Desert 135–8, 145–50, 157–60
Salyut 1 space station 277
Salyut 4 space station 269
Samburu 164
sand dunes 136
sandstorms 136, 137, 156
Savinykh, Viktor 273
sawgrass 59
schistosomiasis 40–1
Schlitz, Lucien 92, 96–7, 98, 99
scopolamine 101, 275
scorpions 84–5
sea lions 171–2, 173
seabirds 114, 115, 116, 159
seals 178–9
SeaOrbiter 199–200
seasickness 99–100, 101
seawater, drinking 117–18, 120, 127
Sebastián (Amazon Indian) 28, 29, 30, 36, 40, 47, 84, 334
serotonin 76, 79
sharks 101, 107–9
Sherpas 1, 4, 16, 210, 224, 229, 231–2, 234, 236, 247, 254, 314, 316
shivering 218, 220, 222
Sinai 150
sinuses 177
skin pigmentation 164–5
skull fractures 3, 129
sleep–wake cycle 281
sleeping bags 247
snakes 45–52, 58–9
  snakebites, treating 50–2
snow blindness 250, 254–5
solar flares 286, 287, 288
solar radiation 139, 265, 286
solar still 102, 103
soldier ants 63–4
Sourbeer, Jay 262
Soyuz 11 mission 277
Soyuz 18 spacecraft 270
space shuttles 287
space sickness 274–5
space stations 275–7, 287
space suits 278–9, 280
space walks 278, 279–80, 297
spacecraft 259–60
  docking 275, 297
  face cameras 294–5
  life on board 281–2, 289–96
  workouts 293–4
spearfishing 103–4, 109
spiders 85, 176
spleen 214
starches 111–12
starvation 122, 123
stingrays 44, 73
Strekalov, Gennadi 269–70
stress 153, 294
stress proteins 153
stroke victims 311
sugars 111
suicide 28, 35, 94, 125
sulphur powder 59
surgical instruments 29–30, 216
survival instinct 6, 7, 16, 57, 70, 89, 105, 123, 148, 221, 302, 303, 304–8, 311, 313, 320
sweating 141, 142, 143, 146, 151–2, 165
Sweigert, Jack 282, 283
synovial fluid 276

taboos, breaking 123, 125
Talacek, James 263
Taqurahu 9, 10
tarantulas 85
telemedicine 18
temperatures
  deserts 137, 157
  outer space 265
  see also body temperature

tendons 33, 34
termite nests 82
termitophiles 38
testicles 163–4
thalamus 309, 312, 314, 315
thermoreceptors 140
thinking 241, 310–11, 316, 317, 318–19
thirst 99, 102, 117, 143, 144, 149
Tierra del Fuego 221
tigers 14–15
Titanic 98
Titov, Vladimir 269–70
toads 163
Todd, Bill 262
toothache 234
tourniquet 32, 33, 35
tribal wisdom 88
tryptamine 79
Tsibliyev, Vasily 275, 277
Tuareg people 137, 150, 159, 160, 163
tungas 54–5
turtles 107, 114, 116

undersea mineral springs 183–4, 186
unmanned deep-sea submersibles 194
urinary infections 73
urine 142, 144, 152
  drinking 149, 158
  salts 158
Uruguayan rugby team 124–5, 313

vaccines 52, 151
Valencia, Pablo 165, 166–70, 190, 302, 307, 317, 318
Valsalva manoeuvre 178
Van Allen Belts 286, 287
Van Pham, Richard 129–30
venom
  conga ants 62
  poison-dart frogs 65, 72, 74, 81, 89
  scorpions 85
  snakes 48–9, 50, 51, 52
  stingrays 44
Viesturs, Ed 250, 254
viridis 79
Virola 79
visions 78
vomiting 100

wadis 150, 158, 162
Walter Reed Hospitals 17
wasps 85
water pressure 177, 181, 185, 193
water spiders 176
Watson, Terry 130
Weathers, Beck 190, 244, 247, 248, 250, 254, 255–7, 258, 302, 307, 308, 317, 318, 319–20
weight loss 122, 123, 131, 150, 167
weightlessness 271, 274, 277, 285, 292, 297
whales 91, 106, 110, 127
Whitson, Peggy 264
will to survive *see* survival instinct
wind shear 92
winds, desert 136, 138
witch doctors 42, 69

Yaga Indians 221–2
yaks 233
Yale University 17
yoga 316

Zancudo Cocha 38, 51, 56